Quantum Mindfulness: Navigating Your Inner Dimensions

Harmony Weaver

Introduction

Welcome fellow explorers,

I invite you to join me on an extraordinary expedition into the uncharted territories of consciousness. As we begin this transformative mission together, I am excited to share with you the revelations and insights that have shaped my understanding of mindfulness, self-discovery, and the boundless capacities of the human mind.

The genesis for this book is rooted in my fascination with the potential for advancing human consciousness and energy fields, and how that is related to discoveries in quantum mechanics. As you unravel the enigmas of the quantum field, a discernible pattern unfolds, a fundamental connection between the

functioning of the quantum world and the intricate workings of our consciousness.

"Quantum Mindfulness: Navigating Your Inner Dimensions" is not just a guidebook; it's a personal narrative woven with the threads of my own quest for understanding. It draws upon the delicate balance between science and spirituality, offering practical insights for those seeking to harmonize the complexities of modern life with a mindful existence.

In a world that often places emphasis on external validations I have decided to go a different route and investigate the dimension of self-exploration. It's given me the space to cultivate a profound relationship with my own thoughts, emotions, and the infinite possibilities that lie within.

I have come to the realization that assisting people in reaching their goals is my calling. A calling to facilitate positive transformations, guide others on their journey to self-mastery, and illuminate the path toward a more purposeful and mindful existence.

In the pages that follow, we'll unravel the mysteries of the mind, explore the layers of consciousness, and discover the power we hold within ourselves. This is an invitation to embrace the principles of quantum mindfulness, a voyage into the heart of our being, where the quantum and the mindful converge in a harmonious dance. Woven into the narrative are tangible tools and real-life examples designed to empower you in the application of quantum mindfulness. From exercises that facilitate a mindful quantum leap to practical techniques for navigating the intricacies of inner dimensions, I will offer hands-on tools that can be seamlessly integrated into your daily life. Whether you're seeking harmony in your thoughts, cultivating a deeper connection with the present moment, or aiming for a mindful quantum shift, these practical applications will guide you towards a more mindful and purposeful existence. Get ready to explore, experiment, and become armed with the tools to enrich your life and awaken the quantum mindfulness within.

May this book serve as a lantern, casting light on your path of self-discovery and growth. Together, let's navigate the inner dimensions of our consciousness and unlock the incredible potential that awaits within.

With gratitude,

Harmony Weaver

Chapter 1: Foundations of Quantum Mindfulness

It is essential to lay down the foundations that bridge the profound principles of quantum mechanics with the art of mindfulness. In this chapter, we will investigate the symbiotic relationship between these seemingly disparate worlds, recognizing their inherent connection and the potential they hold for personal transformation.

The Dance of Particles and Perception

At the heart of the quantum world lies a mesmerizing dance of particles, a ballet choreographed by the laws of quantum mechanics. Envision an expanse where particles exist in states of superposition, simultaneously occupying multiple states until observed. This dance of uncertainty and interconnectedness mirrors the ever-shifting nature of our thoughts and perceptions. Take for example the mysteries of quantum entanglement, where particles, once entwined, instantaneously influence each other's states regardless of distance. In mirroring this entanglement, we see the interconnected web of human consciousness. The principles of quantum superposition and entanglement extend a metaphorical hand, inviting us to reflect on the interdependence of our thoughts, emotions, and the intricate dance of our own perception.

In contemplating the wave-particle duality, we see parallels with the duality inherent in human experience. Just as light can exist as both a wave and a particle, our thoughts oscillate between the tangible and intangible realms of consciousness. This duality extends beyond the quantum realm, resonating with the dichotomy of our own existence; the dance between the physical and the metaphysical, the seen and the unseen. Through this exploration, we set the stage for understanding the intricate relationship between the dance of particles at the quantum level and the dance of our thoughts in the vast landscapes of our minds. This connection serves as the cornerstone for grasping the transformative potential of quantum mindfulness, where the principles of the quantum world become metaphors for the exploration of our inner dimensions.

Mindfulness as Observer and Observed

Key to grasping the essence of quantum mindfulness is recognizing the role of the observer. There is a parallel between the act of observation in quantum physics and the mindful awareness of our thoughts and emotions. As we continue to navigate this terrain, we will uncover how our consciousness shapes our reality, and in turn, how mindfulness can influence the unfolding dance of the quantum realm within. In the quantum realm, the observer is not a passive bystander but an active participant shaping the reality of particles through the act of observation. Similarly, in the landscape of mindfulness, we recognize the profound role of the observer; ourselves. Through deliberate and non-judgmental awareness, we become the architects of our inner reality. As we journey into this fusion of quantum principles and mindfulness, we will understand the dynamic interplay between the observed and the observer.

Drawing inspiration from the renowned double-slit experiment, where particles exhibit both wave and particle characteristics depending on whether they are observed, we begin to discern the parallels with the nature of our thoughts. The observer effect, in both quantum physics and mindfulness, underscores the transformative power of conscious awareness. By acknowledging the impact of our attention on the unfolding moments of our lives, we step into the realm of intentional living.

Consider the observer in the quantum world, influencing the outcome merely by the act of observation. Similarly, our mindful awareness shapes the contours of our inner landscape. Through mindfulness practices, we cultivate the ability to witness our thoughts without undue attachment, allowing them to flow through the canvas of our consciousness without defining our reality. This mindful observation becomes a powerful catalyst for change, as we learn to navigate the delicate dance between being the observed and the observer in the theater of our minds.

As we navigate the intricacies of this relationship, we encounter the transformative potential of mindfulness, the ability to choose our responses rather than reacting impulsively. Through practical exercises, I invite you to embrace the role of the mindful observer, recognizing the agency it possesses in shaping your own narrative. This intentional awareness becomes a guiding principle, steering us towards a state of heightened consciousness where the observer and the observed engage in a dance that transcends the limitations of conventional perception.

Entering the Quantum Field of Now

Central to mindfulness is the art of being present, and we can draw connections to the concept of the "now" in quantum physics. The quantum field is a web of infinite possibilities converging in the present moment. Practical exercises will guide us in entering this quantum field of now, inviting us to experience the richness of the present with heightened awareness. Step with me into the fascinating concept of "now," a fundamental aspect of both quantum mechanics and mindfulness philosophy. In the quantum world, the concept of the "now" takes on a unique significance. The Uncertainty Principle teaches us that the more precisely we know a particle's position, the less precisely we can know its momentum, and vice versa. This inherent uncertainty in the quantum realm emphasizes the crucial role of the present moment, as the act of measurement itself influences the outcome.

In our exploration of mindfulness, the "now" becomes a gateway to profound experiences. Mindfulness encourages us to anchor ourselves in the present, fostering a heightened awareness of the current moment. This concept aligns seamlessly with the quantum understanding that the "now" is not merely a fleeting instant but a dynamic space where possibilities converge. Through mindfulness practices, we learn to be fully present, not dwelling on the past or anticipating the future, but engaging wholeheartedly with the richness of the current moment.

Inspired by the Copenhagen interpretation, we understand the idea that the quantum field is a matrix of infinite possibilities collapsing into a defined state at the moment of observation. Similarly, as we immerse ourselves in the practice of mindfulness, we witness the transformative power of the present moment. By consciously inhabiting the "now," we open ourselves to a realm of possibilities, unburdened by the weight of past regrets or the anxiety of an uncertain future.

Practical exercises accompany our exploration, offering tangible methods for entering the quantum field of now. Guided meditations and mindfulness techniques become gateways, inviting us to experience the expansiveness of the present moment. Let the wisdom of the quantum "now" intertwine with the mindfulness philosophy, paving the way for a deeper understanding of the interconnected dance between our awareness and the infinite possibilities that unfold within the quantum field of the present.

Principles of Quantum Mindfulness

Building on these foundational explorations, we identify key principles that form the bedrock of Quantum Mindfulness. From non-locality to the role of intention in shaping our reality, we connect the dots between quantum concepts and mindfulness practices. These principles serve as guiding lights, illuminating the path towards a more conscious and intentional way of living. Quantum physics introduces us to the concept of non-locality, where particles can instantaneously influence each other regardless of distance. In Quantum Mindfulness, we adopt a similar perspective, recognizing the interconnected nature of our thoughts and emotions. The principle of non-locality invites us to transcend the limitations of space and time in our awareness, fostering a sense of interconnectedness with ourselves and the world around us. By embracing non-locality, we break free from the confines of isolated thinking and open ourselves to a more holistic understanding of our existence.

In the quantum world, the act of observation is not passive; it is an intentional engagement that influences the outcome. Similarly, in Quantum Mindfulness, intentionality becomes a powerful tool for shaping our reality. By infusing our thoughts, actions, and perceptions with purposeful awareness, we become conscious co-creators of our experiences. This principle empowers us to align our intentions with our actions, fostering a harmonious resonance between our inner aspirations and the external manifestations of our lives.

The Uncertainty Principle teaches us that there are inherent limits to the precision with which we can know certain pairs of properties in the quantum world. In Quantum Mindfulness, we recognize the beauty of uncertainty as an invitation to embrace the fluidity of our experiences. Rather than seeking rigid control, we learn to navigate the ebb and flow of life with grace and adaptability. This principle encourages us to cultivate a mindset that thrives in the face of uncertainty, finding resilience and growth in the midst of life's unpredictable nature.

Just as entangled particles influence each other regardless of separation, Quantum Mindfulness underscores the principle of interconnectedness. We acknowledge the intricate web of relationships, within ourselves, with others, and with the universe. By recognizing our interdependence, we cultivate compassion, empathy, and a deep sense of unity. This principle encourages us to approach life with a heart open to the interconnected tapestry of existence, fostering a mindful awareness of the impact our thoughts and actions have on the greater whole.

In embracing these principles, we recognize we are on a transformative path guided by the wisdom of both quantum mechanics and mindfulness. Each principle offers a unique lens through which we can view our lives, providing a framework for conscious living and self-discovery. As we internalize these principles, we set the stage for a mindful quantum leap, a leap into the boundless possibilities that await within the harmonious convergence of the quantum and the mindful.

Chapter 2: Mindful Quantum Leap

In the vast landscape of our consciousness, I define the Mindful Quantum Leap as a profound journey of intentional transformation and heightened awareness. This chapter serves as a practical guide, offering techniques for making a conscious shift in perception and exercises designed to enhance mindful living. As you navigate this exploration, keep in mind that the subsequent chapters will unveil even more practical exercises, expanding your toolkit for intentional awareness. Prepare to discover your higher self, and let the principles of quantum mindfulness become transformative tools for navigating the ever-changing terrain of the mind.

Central to the Mindful Quantum Leap is the art of conscious perception, a deliberate choice to see the world through the lens of mindfulness. In this section, we explore techniques that empower you to observe your thoughts without judgment, fostering a heightened awareness of the narratives unfolding within your mind. Through mindfulness exercises, you'll learn to shift from automatic reactions to intentional responses, laying the foundation for a more conscious and purposeful existence.

Quantum Breathwork: Harmonizing Breath and Consciousness

The breath emerges as a sacred bridge connecting the tangible with the intangible, the physical with the metaphysical. Quantum Breathwork is a profound technique that invites you to harness the transformative power of your breath, aligning it with the rhythmic dance of conscious awareness. As you start the process of harmonizing breath and consciousness, prepare to experience a Mindful Quantum Leap like never before.

The foundation of Quantum Breathwork lies in recognizing the inherent rhythm of your breath and its resonance with the quantum field. In this section, we explore the cadence of breath as a reflection of the ebb and flow of the quantum waves. Through guided practices, you'll attune your awareness to the subtle nuances of each inhale and exhale, forging a profound connection between your breath and the quantum dance of possibilities.

Conscious Inhalation: Infusing Energy and Presence

Conscious inhalation becomes a gateway to infuse your being with vitality and presence. Through specific breath techniques, you'll learn to draw in the energy of the present moment with each breath. As you inhale, envision the quantum field of possibilities converging within you, filling your being with a sense of aliveness and heightened awareness. This practice becomes a conscious communion with the quantum realm, allowing you to tap into the infinite potential that resides within the breath.

Mindful Exhalation: Releasing Tension and Surrendering

The exhale becomes a sacred act of releasing tension and surrendering to the present moment. Quantum Breathwork guides you through mindful exhalation practices, where each breath-out becomes a conscious letting go of thoughts, emotions, and any resistance held within. As you exhale, visualize the release of quantum possibilities into the field around you, creating space for new and transformative energies to enter.

Breath Coherence: Aligning Heart and Mind

Quantum Breathwork extends beyond the physical act of breathing; it becomes a pathway to aligning the heart and mind. Through coherence exercises, synchronize your breath with the natural rhythm of your heart. As the heart and mind harmonize, a state of coherence emerges, creating a resonance that amplifies your connection with the quantum field. This alignment becomes a potent tool for grounding yourself in the present, fostering emotional balance, and enhancing overall well-being.

The culmination of Quantum Breathwork is the Quantum Breath Meditation; an immersive experience that integrates all elements of breath, consciousness, and quantum resonance. This guided meditation takes you on a journey within, where each breath becomes a step into the quantum dimensions of your own consciousness. As you navigate this meditation, you'll tap into the transformative potential of Quantum Breathwork, unlocking a deeper connection with your inner self and the expansive quantum field.

Quantum Breath Meditation: A Journey Within

The Quantum Breath Meditation is a transformative and immersive experience designed to integrate breath, consciousness, and the resonant energy of the quantum field. This guided meditation takes you on a profound journey within, where each breath serves as a portal into the vast and intricate dimensions of your own consciousness. Prepare to embark on this exploration with a sense of openness, curiosity, and a willingness to surrender to the rhythmic dance of your breath.

1. Set the Space: Creating a Sacred Environment

Begin by finding a quiet and comfortable space where you won't be disturbed. Sit or lie down in a relaxed position. Dim the lights, play soft instrumental music, or choose a backdrop that resonates with tranquility. This sets the stage for a sacred and undisturbed meditation experience.

2. Grounding and Centering: Connecting with the Present Moment

Close your eyes and take a few moments to center yourself. Feel the support of the surface beneath you, whether it's a chair, cushion, or the floor. Bring your attention to your breath, inhaling deeply through your nose and exhaling slowly through your mouth. With each breath, let go of any tension or distractions, allowing yourself to fully arrive in the present moment.

3. Conscious Breathing: Attuning to the Rhythmic Dance

Shift your focus to your breath. Inhale deeply, feeling the expansion of your chest and the filling of your lungs. Exhale slowly, releasing any residual tension. Let the breath become a rhythmic dance, a continuous flow of energy moving in and out. As you breathe consciously, visualize each breath as a wave, resonating with the quantum frequencies around you.

4. Quantum Visualization: Stepping into Conscious Dimensions

With each inhalation, visualize that you are stepping into the quantum dimensions of your own consciousness. Picture a gateway opening before you, inviting you to explore the vast landscapes within. As you exhale, imagine releasing any mental chatter or resistance, allowing yourself to enter the quantum field with a clear and open mind.

5. Breath Coherence: Harmonizing Heart and Mind

Integrate breath coherence into the meditation. Sync your breath with the natural rhythm of your heartbeat. Feel the harmonious dance between your heart and breath, creating a coherence that amplifies your connection with the quantum field. This synchronization becomes a bridge between the physical and metaphysical aspects of your being.

6. Quantum Resonance: Feeling the Energy Flow

As you continue breathing consciously, allow yourself to feel the subtle energy flowing through your body. Sense the resonance between your breath, consciousness, and the quantum field. Visualize this energy as vibrant waves, moving in tandem with your breath, creating a symphony of resonance within.

7. Intention Setting: Infusing Breath with Purpose

Bring intentionality to your breath. With each inhalation, set a positive and empowering intention for the journey within. This could be a statement or affirmation that aligns with your deeper aspirations. As you exhale, release any doubts or limitations that may hinder the realization of your intention.

8. Exploring Inner Dimensions: Guided Visualization

Guided by your breath and intention, begin exploring the inner dimensions of your consciousness. Picture a serene landscape or a sacred space within your mind. Allow the quantum dimensions to unfold, revealing insights, memories, or symbols that hold personal significance. Trust the process and let the exploration unfold organically.

9. Quantum Integration: Merging with the Infinite

As the meditation progresses, visualize yourself merging with the infinite quantum field. Sense the interconnectedness of your being with the boundless possibilities surrounding you. With each breath, feel the boundaries of your individual self dissolving, expanding into the quantum tapestry of existence.

10. Closing and Reflecting: Returning to the Present

Gradually bring your awareness back to your physical surroundings. Slowly reintegrate with the present moment, maintaining a sense of gratitude for the journey within. Take a few deep breaths, and when you're ready, gently open your eyes.

This Quantum Breath Meditation serves as a powerful tool for self-discovery and a mindful quantum leap. Regular practice can deepen your connection with the quantum field, enhance your conscious awareness, and open doors to profound insights within the realms of your own consciousness.

Begin the exploration of Quantum Breathwork with an open heart and a willingness to surrender to the rhythmic dance of your breath. This practice is a gateway to profound self-discovery and a mindful quantum leap, where the breath becomes a vehicle for traversing the landscapes of the quantum and the mindful in perfect harmony.

Meditation becomes a quantum exploration as we introduce the Quantum Observation Meditation. This practice invites you to observe your thoughts as if they were particles in the quantum field. By cultivating a detached yet engaged awareness, you'll learn to navigate the currents of your consciousness with grace and clarity. This meditation serves as a compass, guiding you through the intricate landscapes of your mind.

Quantum Observation Meditation: A Deep Dive into Conscious Awareness

The Quantum Observation Meditation is a transformative practice that merges principles of quantum mechanics with mindfulness techniques. This guided meditation encourages you to observe your thoughts as if they were particles in the quantum field, cultivating a detached yet engaged awareness. By navigating the intricate landscapes of your consciousness, this meditation serves as a compass, guiding you through the quantum dimensions of your own mind. Prepare for a journey of self-discovery and heightened awareness with the Quantum Observation Meditation.

1. Set the Stage: Creating a Calm Atmosphere

Find a quiet and comfortable space for your meditation. Ensure that you won't be disturbed. Sit or lie down in a relaxed position. Dim the lights, play soft ambient music, or choose an environment that promotes tranquility. The aim is to create a space conducive to deep introspection and focus.

2. Grounding and Centering: Establishing Connection

Close your eyes and take a few deep breaths. Feel the support of the surface beneath you. Inhale deeply through your nose, and exhale slowly through your mouth. With each breath, let go of any tension or distractions. Bring your attention to the present moment, grounding yourself in the here and now.

3. Breath Awareness: Initiating the Quantum Connection

Shift your focus to your breath. Notice the sensations of inhalation and exhalation. Let your breath become the anchor that connects you to the present moment. As you breathe, envision the quantum field around you, a vast and energetic space waiting to be explored with the power of your conscious awareness.

4. Observer Mindset: Cultivating Detached Engagement

Embrace an observer mindset. As you continue to breathe, adopt the perspective of an observer watching thoughts drift through the quantum landscape of your consciousness. Allow your thoughts to arise without judgment or attachment. Picture them as particles in constant motion, emerging and dissipating within the quantum field.

5. Quantum Visualization: Thoughts as Quantum Particles

Visualize your thoughts taking on the characteristics of quantum particles. Picture them as dynamic and ever-changing entities, existing in a state of flux. Each thought is like a particle in the quantum field, influenced by the observer (you) and subject to the principles of uncertainty and entanglement.

6. Detached Observation: Allowing Thoughts to Flow

Maintain a detached yet engaged stance. As thoughts arise, observe them without getting entangled. Let them flow through your consciousness like particles dancing in the quantum field. Resist the urge to analyze or attach significance to each thought. Instead, remain in a state of serene observation.

7. Intentional Focus: Guiding Your Awareness

While observing thoughts, set an intention for meditation. It could be focused on clarity, inner peace, or gaining insights into a particular aspect of your life. The intention becomes a guiding force, directing your conscious awareness within the quantum field of your thoughts.

8. Quantum Flow: Navigating the Mind's Landscape

Allow your attention to flow with the quantum currents of your mind. Notice how thoughts connect, intersect, and create patterns. Embrace the fluidity of the quantum flow, acknowledging the interconnected nature of your thoughts within the vast quantum landscape.

9. Centering Breath: Returning to the Present Moment

When distractions arise or the mind starts to wander, gently bring your focus back to your breath. The breath serves as a constant anchor, reconnecting you with the present moment and the quantum field of your thoughts. Inhale clarity, exhale distractions.

10. Closing and Reflection: Integrating Insights

As the meditation comes to a close, take a few moments to reflect on the observations made during the practice. Notice any shifts in your awareness or insights gained from observing the quantum dance of your thoughts. Express gratitude for the journey and slowly bring your awareness back to your surroundings.

Conclusion: Embracing Quantum Awareness

Quantum Observation Meditation is a powerful tool for cultivating conscious awareness within the intricate dimensions of your own consciousness. Regular practice of this meditation can deepen your understanding of the interconnected nature of thoughts and emotions, fostering a sense of detached yet engaged mindfulness within the quantum field of your mind.

Mindful Living Exercises: Integrating Awareness into Daily Life

Mindful living extends beyond meditation cushions and into the fabric of our daily experiences. In this section, we provide practical exercises designed to seamlessly integrate mindfulness into your routine. From mindful eating to conscious listening, these exercises offer opportunities for intentional awareness in every aspect of your life. As you engage with these practices, you'll discover the transformative potential of infusing mindfulness into all areas of your daily existence.

Below are practical exercises designed to seamlessly integrate mindfulness into various aspects of your routine. These exercises aim to enhance intentional awareness, bringing mindfulness into everyday activities for a more conscious and purposeful existence.

1. Mindful Eating Exercise: Savory Awareness

Objective: Cultivate awareness and appreciation for the act of eating, savoring each bite mindfully.

Instructions:

Choose a meal or snack and set aside time to eat without distractions.

Before taking the first bite, take a moment to observe the colors, textures, and aromas of the food.

As you lift the utensil or bring the food to your mouth, notice the sensations in your hand and the movement of your arm.

Chew slowly and deliberately. Pay attention to the flavors, textures, and how the food feels in your mouth.

Be present with each bite, and if your mind starts to wander, gently bring it back to the sensory experience of eating.

2. Mindful Walking Exercise: Steps of Awareness

Objective: Transform your daily walk into a mindful exercise, grounding yourself in each step.

Instructions:

As you begin your walk, start by bringing attention to your breath. Inhale and exhale consciously.

Feel the connection between your feet and the ground. Notice the sensation of each step.

Pay attention to the movements of your body—how your arms swing, how your torso shifts with each step.

Engage your senses. Notice the sounds around you, the feel of the air, and any scents in the environment.

If your mind starts to wander, gently redirect your focus to the physical sensations of walking.

3. Mindful Listening Exercise: Sonic Awareness

Objective: Sharpen your listening skills and deepen your connection to the sounds around you.

Instructions:

Find a quiet place to sit or stand.

Close your eyes and focus on the sounds in your environment.

Identify individual sounds, whether they are distant or nearby, soft or loud.

Allow each sound to come and go without judgment or analysis.

Expand your awareness to include the entire auditory landscape.

Notice how your perception of sound changes as you become more attuned to the present moment.

4. Mindful Breathing Exercise: Centering Breath

Objective: Use intentional breathing to bring yourself back to the present moment and cultivate a sense of calm.

Instructions:

Find a comfortable seated position.

Close your eyes and bring attention to your breath.

Inhale deeply through your nose, feeling the breath fill your lungs.
Exhale slowly through your mouth, releasing tension with each breath out.
Focus on the rise and fall of your chest or the sensation of air passing through your nostrils.
If your mind begins to wander, gently guide it back to the rhythmic flow of your breath.

5. Mindful Daily Routine Exercise: Ritual Awareness

Objective: Infuse mindfulness into your daily routine by bringing attention to each task.

Instructions:

Choose a routine task, such as brushing your teeth, washing dishes, or taking a shower.
Approach the task with intention. Notice the sensations, movements, and details involved.

Engage your senses. Feel the water on your hands, notice the scent of soap, or the taste of toothpaste.

Resist the urge to rush. Perform the task deliberately, savoring each moment.

If your mind starts to wander, gently redirect your focus to the sensory experience of the task at hand.

Note: These exercises can be adapted to fit various daily activities. The key is to approach each task with a heightened sense of awareness and intention, allowing mindfulness to become an integral part of your routine. Regular practice can contribute to a more mindful and purposeful way of living.

Affirmations take on a quantum twist as we explore the power of intentional language to shape our reality. Quantum Affirmations become tools for consciously influencing the quantum field of possibilities. Through guided exercises, you'll learn to craft affirmations that align with your truest aspirations, creating a positive resonance that ripples through the quantum fabric of your life.

Quantum Affirmations: Shaping Reality with Intentional Language

Quantum Affirmations serve as potent tools for consciously influencing the quantum field of possibilities. By infusing intentional language with positive energy, you can shape your reality and align your thoughts with your truest aspirations. This practice goes beyond traditional affirmations, incorporating the principles of quantum mechanics to enhance their transformative potential. Explore the following steps to harness the full power of Quantum Affirmations and empower yourself on your journey of mindful living.

1. Setting Intentions: Clarifying Your Desires

Before crafting Quantum Affirmations, take a moment to reflect on your desires and intentions. What aspects of your life do you wish to enhance or transform? Whether it's fostering self-love, improving relationships, or achieving personal growth, clearly define your intentions to provide a solid foundation for your affirmations.

2. Positive Framing: Emphasizing the Positive

Frame your affirmations in a positive and empowering manner. Focus on what you want to manifest rather than what you wish to avoid. For example, instead of saying, "I will not be stressed," rephrase it as "I am cultivating inner peace and resilience." This positive framing directs your energy towards constructive outcomes.

3. Present Tense: Anchoring in the Now

Phrase your affirmations in the present tense to create a sense of immediacy. By expressing your intentions as if they are already happening, you anchor your affirmations in the present moment, aligning them with the principles of the quantum field where the "now" holds immense creative power.

4. Clarity and Specificity: Fine-Tuning Your Affirmations

Ensure that your affirmations are clear and specific. Vague affirmations may lead to ambiguous results. For instance, instead of a general affirmation like "I am successful," specify the areas of success you wish to manifest, such as "I am achieving success in my career, relationships, and personal growth."

5. Emotional Resonance: Infusing Energy and Feeling

As you craft your affirmations, infuse them with emotional resonance. Connect with the feelings associated with achieving your desires. Feel the joy, fulfillment, and gratitude as if your affirmations are already a reality. This emotional charge adds vibrancy to your intentions and enhances their impact on the quantum field.

6. Consistent Repetition: Programming the Subconscious

Repetition is key to embedding affirmations into your subconscious mind. Regularly repeat your Quantum Affirmations with conviction and belief. Create a routine for reciting them, incorporating them into your morning or evening rituals. Consistent repetition strengthens the neural pathways associated with your affirmations, reinforcing their influence on your thoughts and actions.

7. Visualization: Enhancing the Quantum Experience

Combine your affirmations with visualization to amplify their impact. Close your eyes and vividly picture yourself living the reality described in your affirmations. Engage all your senses to create a detailed mental image. Visualization enhances the quantum experience, making your affirmations a vivid part of your consciousness.

8. Gratitude and Acceptance: Embracing the Now

Express gratitude for the present moment and acceptance of the journey toward your desires. Acknowledge the progress you've made and the positive changes occurring in your life. This attitude of gratitude aligns with the quantum principle of appreciating the now and opens the pathway for more abundance and fulfillment.

9. Affirmation Integration: Daily Application

Integrate your Quantum Affirmations into your daily life. Use them as a guiding mantra during meditation, repeat them silently during moments of reflection, or post them in places where you'll see them frequently. Make your affirmations an integral part of your consciousness, allowing them to shape your thoughts and actions throughout the day.

10. Adaptation and Evolution: Embracing Change

Review and adapt your affirmations as your intentions evolve. Life is dynamic, and your desires may change over time. Regularly reassess and refine your affirmations to ensure they align with your current aspirations. Embrace the fluidity of change and allow your affirmations to evolve alongside your personal growth.

By applying these comprehensive steps, Quantum Affirmations can become a dynamic force in shaping your reality and fostering intentional awareness in every aspect of your life. Engage with these affirmations not only as statements but as gateways to a mindful quantum experience, propelling you toward the abundant and purposeful existence you envision.

Take this Mindful Quantum Leap with an open heart and a curious mind. The techniques and exercises within this chapter are not mere practices; they are gateways to a heightened state of awareness, inviting you to dance with the quantum and the mindful in a harmonious rhythm. May this journey lead you to new dimensions of self-discovery and mindful living, propelling you toward the transformative quantum leap that awaits within.

Chapter 3: Navigating Your Inner Dimensions

This chapter serves as a guide for understanding the multifaceted nature of our inner dimensions. We'll unravel the mysteries of consciousness and equip ourselves with transformative tools for navigating the intricate landscapes within. Get ready to engage in self-actualization, where the layers of your consciousness unfold like chapters in a cosmic novel.

Unveiling Layers of Consciousness

The Surface of Awareness:

At the surface of our awareness lies everyday consciousness, a dynamic and ever-present state that shapes our interactions with the external world and influences the fabric of our daily experiences. This layer of consciousness is like the visible spectrum of light, illuminating the tangible aspects of our lives. As we navigate through the demands of daily routines, engage in social interactions, and respond to external stimuli, everyday consciousness becomes the lens through which we interpret and engage with the world.

Everyday consciousness is deeply intertwined with the routine activities that constitute our daily lives. From waking up in the morning to engaging in work, socializing, and winding down at night, our routines are guided by the autopilot of everyday consciousness.

The five senses play a pivotal role in shaping everyday consciousness. Sight, sound, touch, taste, and smell collectively contribute to our perception of the external world. The sensory input we receive forms the basis for how we interpret and navigate our surroundings. Everyday consciousness is characterized by immediate awareness, focusing on the present moment and the tasks at hand. It is the aspect of consciousness that helps us respond to the demands of our environment promptly.

In this state, there is a strong identification with the ego; a sense of self that is anchored in personal identity, roles, and responsibilities. Everyday consciousness often operates within the framework of individuality, emphasizing personal experiences and perspectives. The pragmatic nature of everyday consciousness enables problem-solving and decision-making. It involves analytical thinking, planning, and strategizing to navigate challenges and accomplish daily goals.

Everyday consciousness creates a sense of familiarity and predictability in our lives. It forms the structure within which we operate, fostering a sense of stability and order. This layer of consciousness allows us to adapt to external demands efficiently. Whether at work, in relationships, or facing various responsibilities, everyday consciousness facilitates our ability to respond to the dynamic nature of daily life.

Habits, routines, and automatic behaviors are deeply ingrained in everyday consciousness. When we repeat certain actions, they become habitual, freeing up cognitive resources for more complex aspects of daily living. Everyday consciousness plays a crucial role in social interactions. It involves understanding social cues, interpreting verbal and non-verbal communication, and adapting behavior to fit social norms.

The state of everyday consciousness influences how we manage and express emotions. It involves the immediate processing of emotional stimuli and the regulation of emotional responses in various situations.

While everyday consciousness provides practical tools for navigating the external world, it is just the surface of a vast ocean of consciousness.

Understanding and exploring the deeper layers within—such as the subconscious and collective unconscious—opens the door to a more profound self-discovery and a richer tapestry of human experience. In the chapters ahead, we'll delve into these layers, uncovering the hidden realms that shape the essence of our being.

Subconscious Realms

Unearthing Hidden Influences:
Beneath the surface, the subconscious realms hold the imprints of past experiences, beliefs, and emotions. A reservoir of hidden influences that shape our behaviors, values, and perceptions. Like an iceberg, the subconscious holds the majority of our mental processes below the surface, exerting a profound impact on our thoughts, emotions, and actions. Unraveling the secrets of the subconscious mind unveils the keys to self-discovery, personal growth, and the transformation of deeply ingrained patterns.

The subconscious serves as a vast repository of memories, encompassing experiences, emotions, and sensations from our past. These memories, often beyond conscious recall, contribute to the formation of our identity and influence present-day reactions. Operating beyond the immediate awareness of everyday consciousness, the subconscious is responsible for automatic responses and reactions. These can range from instinctual survival responses to learned behaviors and habits.

The subconscious plays a pivotal role in the formation and maintenance of beliefs. Early life experiences, cultural influences, and societal conditioning contribute to the development of belief systems that operate beneath the surface of conscious thought. Emotions, particularly those tied to past experiences, are processed within the subconscious. Traumas, joys, and unresolved emotions from the past may subtly impact present emotional responses, influencing moods and perspectives.

The subconscious communicates through symbols, metaphors, and imagery in dreams and creative expressions. It operates in a realm of symbolic language that may hold insights into deeper aspects of the self. Engage in self-reflection and journaling to bring hidden influences to light. Explore your thoughts, emotions, and recurring patterns. As you explore the pages of your journal, you may uncover connections between past experiences and present behavior.

Mindfulness allows you to observe the subtle movements within the subconscious realms. Through mindfulness meditation and awareness exercises, you can gently bring hidden thoughts and emotions to the surface, fostering a deeper understanding of the self. Dreams are windows into the subconscious. Keeping a dream journal and analyzing recurring themes, symbols, or emotions in your dreams can reveal hidden influences and unresolved aspects of your psyche.

Professional therapy, such as psychoanalysis or cognitive-behavioral therapy, provides a structured environment for unearthing hidden influences. Therapists guide individuals in exploring deep-seated beliefs, past traumas, and automatic responses that may be impacting their lives.

Creative outlets, such as art, writing, or music, offer a medium for the subconscious to express itself. Engaging in artistic endeavors can unveil hidden influences and provide a channel for catharsis and self-discovery.

Techniques like hypnosis and guided imagery facilitate direct access to the subconscious. Under the guidance of a trained professional, individuals can explore hidden influences, reframe beliefs, and address unresolved issues embedded in the subconscious realms.

Fine tune the core beliefs that govern your thoughts and behaviors. Identify beliefs instilled during childhood or formed through significant life experiences. Questioning and challenging these beliefs can lead to profound shifts in the subconscious landscape.

Understanding the subconscious is a journey of self-exploration and excavation. By unearthing hidden influences, we gain the power to transform ingrained patterns, heal unresolved wounds, and cultivate a more authentic and conscious way of being. As we navigate the depths within, the subconscious becomes a landscape for personal growth and the emergence of a more empowered self.

The Collective Unconscious

Connecting to Shared Archetypes:

Understanding the collective unconscious unveils the threads that connect us to the tapestry of human experience. This layer of consciousness transcends the individual and taps into universal symbols and patterns that have shaped human experiences across cultures and epochs. Exploring the collective unconscious opens a gateway to shared human narratives and archetypal motifs that influence our dreams, myths, and cultural expressions.

Archetypes are innate, universal symbols residing in the collective unconscious. They represent fundamental human experiences and themes that have persisted throughout history. Examples include the Hero, the Mother, the Shadow, and the Wise Old Man.

The collective unconscious transcends cultural boundaries, weaving a tapestry of shared symbols that resonate across diverse societies. Archetypes manifest in myths, folklore, and cultural narratives, offering a common ground for human understanding. While archetypes are universal, their expression is both personal and cultural. Individuals may encounter archetypes in personal dreams, while societies weave them into the fabric of cultural storytelling, rituals, and traditions.

The collective unconscious significantly influences the symbolic language of dreams. Archetypal figures and motifs often appear in dreams, providing individuals with a symbolic reservoir to navigate and understand their own experiences. The Hero's Journey, the Anima/Animus, and the Shadow are examples of archetypal structures that recur in myths and dreams across cultures.

Pay attention to recurring symbols, figures, or themes in your dreams. These may be manifestations of archetypes providing insights into your personal and collective unconscious. Keep a dream journal to analyze and understand these symbols. Explore myths, legends, and folklore from various cultures. Identify archetypal characters and motifs that reappear across different stories. Recognizing these shared elements fosters a deeper connection to the collective unconscious.

Engage in creative activities to access and express archetypal energies. Writing, art, music, and dance can serve as channels for archetypal exploration. Allow the symbols and themes that emerge to guide your creative process. Incorporate meditative practices or guided imagery to explore the depths of your inner landscapes. Visualization exercises can lead to encounters with archetypal figures, providing personal insights and connections to shared human experiences.

Consider exploring analytical psychology through books, workshops, or discussions. Studying archetypal theory and the collective unconscious can deepen your understanding of the psychological and spiritual dimensions of archetypes. Incorporate rituals or symbolic acts into your life that resonate with archetypal themes. These acts can range from personal rituals to communal ceremonies, reinforcing a connection to shared symbols and the collective unconscious.

Embrace the process of shadow work. The shadow concept, represents the unconscious and often repressed aspects of an individual's personality. It encompasses elements such as desires, traits, and emotions that are typically deemed socially unacceptable or personally uncomfortable. Engaging with the shadow involves acknowledging and integrating these hidden aspects, leading to greater self-awareness and personal growth. By confronting and embracing the shadow, you can achieve a more holistic understanding of yourself and cultivate a sense of inner balance.

Connecting to the collective unconscious offers a bridge between the personal and the universal, providing a rich source of symbolism and meaning. By exploring shared archetypes, you tap into the deep currents of human experience, fostering a sense of interconnectedness and enriching our understanding of the human psyche. As you navigate the collective unconscious, you set off on a journey that transcends time and culture, unveiling the timeless threads that weave us into the fabric of humanity.

Transcendent Dimensions:

Tapping into Higher Consciousness
Beyond the individual and collective layers of consciousness lies the awe-inspiring web of transcendent dimensions, a space where you can tap into higher consciousness and connect with the universal and infinite aspects of existence. This layer extends beyond the boundaries of personal identity and cultural conditioning, offering a profound sense of interconnectedness, spiritual insight, and expanded awareness.

Transcendent dimensions encompass a sense of universal connection, where you feel linked to a greater whole. This connection goes beyond individuality, embracing the oneness that underlies all of existence. Tapping into higher consciousness often leads to profound spiritual insights. You may experience moments of clarity, heightened intuition, and a deep understanding of the interconnected nature of all life.

The awareness experienced in transcendent dimensions transcends the limitations of ordinary consciousness. It involves a heightened state of perception, where you may sense the timeless, the boundless, and the infinite. Transcendent dimensions are often associated with mystical experiences, which involve a direct and personal encounter with the divine, the sacred, or the infinite. These experiences can be transformative and transcend the ordinary boundaries of perception. In the dimension of higher consciousness, time takes on a different quality. You may report a sense of timelessness, where past, present, and future converge into a singular, eternal moment.

Meditation serves as a powerful gateway to transcendent dimensions. Techniques such as mindfulness meditation, transcendental meditation, and contemplative practices can facilitate a connection to higher states of consciousness. Engaging in contemplative practices and reflective activities allows you to go beyond surface-level thinking. Journaling, philosophical inquiry, and deep introspection create pathways to transcendent insights.

Immersing oneself in nature can provide a direct experience of transcendent dimensions. The beauty and harmony found in natural settings often evoke a sense of awe and interconnectedness, fostering a connection to higher consciousness.

Certain mind-altering practices, when approached responsibly and intentionally, have the potential to offer glimpses into transcendent dimensions. These practices may include the use of psychedelics, participation in sacred plant ceremonies, or engagement in other consciousness-expanding rituals. It is crucial to prioritize safety and well-being. Before considering any of these practices, it is highly recommended to consult with a qualified healthcare professional or doctor to ensure that they are suitable for your individual health and circumstances.

Engaging in spiritual disciplines, whether through organized religions or personal belief systems, can lead to experiences of higher consciousness. Practices such as prayer, chanting, or rituals provide avenues for transcendent connection.

Some individuals report transcendent experiences during near-death encounters.

These profound episodes often involve a sense of leaving the body, encountering a light, and gaining insights into the nature of existence.

Adopting holistic approaches to well-being, such as yoga, tai chi, or energy work, can align the body, mind, and spirit. These practices create a conducive environment for tapping into higher consciousness.

Tapping into higher consciousness is a deeply personal and transformative journey that transcends the boundaries of ordinary perception. As you explore these transcendent dimensions, they often discover a profound sense of unity, purpose, and the boundless potential that lies within the vast and mysterious realms of higher consciousness.

Tools for Exploring Inner Dimensions

Venturing into the inner dimensions of consciousness requires a set of transformative tools that facilitate self-discovery, mindfulness, and personal growth. These tools serve as guides, helping you navigate the intricate landscapes of your inner selves. From ancient practices to modern approaches, these tools empower you to explore the depths of consciousness and unlock the potential for profound transformation.

Meditation stands as a cornerstone for exploring inner dimensions. Whether through mindfulness, transcendental meditation, or loving-kindness practices, meditation provides a sanctuary for turning inward. By quieting the mind and observing thoughts without attachment, you can access deeper layers of consciousness, fostering inner peace and insight.

Journaling and self-reflection act as mirrors to the soul, offering a method to map the inner landscape. Through the act of putting pen to paper, you can explore thoughts, emotions, and experiences. Journaling becomes a sacred space for self-expression, uncovering patterns, and gaining clarity on the nuances of your inner world.

Dream analysis unveils the symbolic language of the unconscious. Keeping a dream journal and exploring the imagery, emotions, and narratives within dreams provide valuable insights into the hidden realms of the psyche. By decoding these messages, you gain a deeper understanding of your inner world.

Mindful practices in daily life bridge the gap between routine activities and heightened awareness. From mindful eating to conscious listening, you can infuse intentionality into every moment. These practices cultivate presence, allowing you to explore inner dimensions amidst the tapestry of everyday experiences.

Creative expression becomes a dynamic tool for inner exploration. Through art, writing, music, dance, or any form of artistic endeavor, you can tap into the depths of your emotions and thoughts. Creative expression serves as a medium for catharsis, self-discovery, and the unfolding of the soul's unique voice.

Participating in meditation retreats and workshops offers immersive experiences guided by seasoned practitioners. These settings provide a supportive environment for deepening meditation practices, fostering self-inquiry, and connecting with like-minded individuals on a shared journey of inner exploration.

Holotropic Breathwork, a powerful technique, involves controlled breathing to induce altered states of consciousness. This method facilitates deep exploration of the psyche, allowing you to access transformative experiences and navigate inner dimensions with heightened awareness.

Psychotherapy and inner work with trained professionals offer structured guidance for exploring inner dimensions. Therapists help individuals navigate subconscious realms, uncover hidden influences, and address unresolved aspects of the self, fostering healing and personal growth.

Mind-body practices such as yoga, tai chi, and qigong integrate physical movement with mindful awareness. These practices serve as tools for exploring the mind-body connection, promoting holistic well-being, and unlocking the interconnected nature of inner dimensions.

Rituals and ceremonies create sacred spaces for inner exploration. Whether rooted in spiritual traditions or personal symbolism, these intentional practices facilitate connection with the sacred aspects of life, opening gateways to transcendent dimensions within.

Armed with these diverse tools, you can set off on a transformative journey into your inner dimensions. Each tool offers a unique perspective, inviting you to uncover the mysteries of your consciousness, embrace personal growth, and cultivate a more mindful and authentic way of being. It's worth noting that the practices mentioned earlier will be explored in greater depth later on in this book. The integration of these tools forms a comprehensive approach to exploring the rich tapestry of inner dimensions that shape the essence of our existence.

Chapter 4: Bridging Science and Spirituality

As we unravel the mysteries of quantum mechanics and its implications for consciousness, we aim to find a harmonious convergence between scientific inquiry and the profound wisdom embedded in spiritual traditions. The synthesis of these seemingly disparate ideas offers a pathway toward a more comprehensive understanding of the nature of existence and the human experience.

Quantum Principles Unveiled

At the heart of quantum mechanics lies the intriguing concept of wave-particle duality. Subatomic particles, such as electrons and photons, exhibit both particle-like and wave-like properties, blurring the boundaries between classical categories.

Enter the concept of superposition, a phenomenon where particles exist in multiple states simultaneously. Unlike classical objects that occupy definite positions, quantum particles can exist in a multitude of states until observed. The famous thought experiment involving a cat in a superposition of life and death vividly illustrates the profound implications of this principle, challenging our intuitive notions of reality and pushing the boundaries of what we consider possible.

In the mysterious phenomenon of entanglement, where particles become intricately linked regardless of distance; entangled particles, such as electrons or photons, instantaneously influence each other's states, defying the constraints of spacetime. This quantum entanglement, reveals the interconnected nature of the quantum world, suggesting a form of communication that transcends our conventional understanding of physical reality.

The foundational Uncertainty Principle asserts that the more precisely we measure one property of a particle (such as its position), the less precisely we can know another property (such as its momentum). Quantum uncertainty challenges the classical notion of a deterministic universe, introducing an inherent limit to our ability to precisely predict the behavior of quantum entities.

Quantum tunneling allows particles to appear on the other side of energy barriers, defying classical expectations. This remarkable principle finds applications in various fields, from explaining nuclear fusion in stars to the operation of transistors in electronic devices.

The dance of possibilities and the inherent uncertainties of the quantum world become increasingly apparent. These principles lay the groundwork for understanding the profound implications of quantum mechanics on consciousness, reality, and the bridging of science with the timeless wisdom found in spiritual practices.

The Observer Effect: Consciousness and Reality

Central to the Observer Effect lies the idea that the act of observation fundamentally alters the behavior of quantum particles. The classic double-slit experiment exemplifies this phenomenon: when unobserved, particles exhibit a wave-particle duality, existing in a superposition of states. However, the moment an observer measures or observes the particles, they "collapse" into a specific state. This shift from potentiality to actuality emphasizes the inseparable link between the observer and the observed, raising profound questions about the nature of reality. In the quantum world, the presence of a conscious observer seems to play a pivotal role in determining the outcome of an experiment. This realization sparks contemplation about the extent to which consciousness itself is entangled with the fabric of reality. Does consciousness shape the world, or does the world shape consciousness? The Observer Effect suggests an intricate dance between the two.

Quantum Entanglement and Non-Locality: Spooky Action at a Distance

When particles become entangled, the measurement of one particle instantaneously influences the state of the other, regardless of the distance between them. This non-local connection challenges classical notions of causality and hints at a form of communication that transcends space and time. The intertwined nature of entanglement and the Observer Effect invites contemplation on the interconnectedness of all things.

The traditional view of an objective, independent reality independent of observation is challenged by the quantum understanding that observation is an integral part of the process. Conscious observership becomes inseparable from the creation and manifestation of reality at the quantum level, prompting a reevaluation of the relationship between consciousness and the external world.

The theory of participatory realism suggests that reality is not an independent, pre-existing entity but is co-created through the active participation of observers. The notion that consciousness plays a fundamental role in shaping reality challenges the classical separation between subject and object, paving the way for a more integrated understanding of the observer-world relationship.

As we navigate the profound implications of the Observer Effect, the inseparable link between consciousness and the quantum world comes into focus. This section invites readers to contemplate the role of conscious awareness in the unfolding dance of quantum reality and sets the stage for bridging the gap between scientific exploration and the spiritual dimensions of consciousness.

Bridging the Abyss: The Marriage of Quantum Physics and Consciousness

Some theories propose that entangled particles not only exhibit instantaneous correlations but might also mirror a deep interconnectivity present in the collective human experience. This perspective invites contemplation on the interconnected nature of consciousness and the quantum fabric of reality.

Consider the notion that consciousness acts as the ultimate quantum observer, influencing the outcomes of quantum events. As particles exist in a superposition of states until observed, the role of consciousness in collapsing this quantum potentiality into actuality becomes a captivating proposition. This viewpoint raises profound questions about the nature of consciousness and its inherent entanglement with the unfolding quantum reality.

The synchronized behavior of particles might find a counterpart in the coherence of human consciousness. Analogous to how entangled particles exhibit correlated states, collective states of focused human consciousness may demonstrate a form of quantum coherence. This synergy causes us to contemplate the intricate relationship between individual and collective consciousness and its potential influence on the quantum world.

Quantum Consciousness: Integrating Mind, Body, and Spirit

If consciousness is entangled with the quantum fabric of the body, can intentional focus and awareness influence physical health? Quantum healing theories propose that by aligning consciousness with positive intentions and thoughts, individuals may contribute to their own well-being. This holistic perspective encourages a deeper exploration of the mind's role in the intricate dance of health and healing.

Some theories propose that consciousness operates as a form of quantum resonance, suggesting that the vibrational frequencies of consciousness harmonize with the quantum fields that permeate the body. This resonance creates a symphony of interconnected frequencies, influencing not only the physical body but also resonating with the larger quantum tapestry of the universe. The concept of quantum resonance invites contemplation on the subtle energies that underlie conscious experience. From the firing of neurons to the regulation of cellular activities, quantum phenomena may play a role in the orchestration of the body's functions. This holistic view challenges the notion of consciousness as a mere byproduct of neural activity, inviting exploration into the quantum underpinnings of embodied awareness.

Chakras and Quantum Energy Centers: Bridging the Energetic Plane

Chakras are often regarded as vital energy hubs, each associated with specific qualities and functions. Similarly, the concept of quantum energy centers suggests the existence of focal points within the quantum body where energy converges and resonates. This convergence beckons individuals to deepen their understanding of these energetic realms, exploring how the subtle dance of energy within the body aligns with the quantum currents that underlie the very essence of existence.

The balanced circulation of energy is considered essential for maintaining physical, emotional, and spiritual equilibrium. This mirrors the quantum understanding that resonant frequencies may influence the quantum field, shaping the dynamic interplay of particles and waves. This convergence encourages a deeper exploration of the subtle energies that underlie conscious experience, inviting individuals to align with the rhythmic dance of life force energy and resonate with the quantum currents that permeate the cosmos.

Sacred Geometry: Quantum Patterns in the Fabric of Creation

Sacred geometry often symbolizes the archetypal patterns that underlie creation, representing the interconnected relationship between the microcosm and macrocosm. Similarly, the concept of quantum patterns elucidates the fundamental structures governing the behavior of particles and waves. This harmonious alignment encourages contemplation on the universal symmetries, proportions, and inherent order present in the quantum dimension.

Eastern Philosophy Meets Quantum Reality

In mindfulness, the emphasis on being present in the now finds resonance with the synchronized behavior of quantum particles in a state of coherence. This meeting of concepts illuminates the transformative power of aligning conscious awareness with the harmonious flow of quantum processes, fostering a deep sense of presence and interconnected being.

Karma suggests a web of interconnected actions and consequences, mirroring the intricate tapestry woven by quantum causality. The recognition that every action influences the quantum fabric of reality invites reflection on the profound responsibility inherent in both karmic philosophy and quantum interconnectedness.

Enter the concept where the emptiness of mind from Eastern meditation traditions converges with the quantum concept of potentiality.

Emptiness in meditation signifies a still mind free from attachments, opening a space for insight and clarity. Quantum potentiality similarly represents a state of infinite possibilities before observation collapses them into actuality. This meeting of principles emphasizes the transformative power of stillness and openness to the quantum field of potentiality.

Yin-Yang represents the interdependence and balance of opposites, a concept mirrored in the wave-particle duality of quantum entities. This union invites contemplation on the dynamic interplay of polarities, where the dance between opposing forces mirrors the intricate interdependence found both in Eastern wisdom and quantum reality.

Metta emphasizes universal compassion and interconnectedness, transcending boundaries. Quantum entanglement, with its non-local connections, suggests a form of universal entanglement that echoes the expansive nature of metta. This meeting of concepts invites a deep exploration of compassion that transcends the limitations of space and time.

In the convergence of Eastern philosophy with the principles of quantum reality, a tapestry of wisdom unfolds; a synthesis that transcends cultural divides, offering profound insights into the nature of consciousness, interconnectedness, and the holistic fabric of existence. This harmonious union invites individuals to draw from the timeless well of Eastern wisdom and the cutting-edge insights of quantum physics, fostering a holistic approach to understanding the mysteries of the universe and the human experience.

Meditation and Quantum Coherence:

In the act of meditation, individuals assume the role of conductors, guiding the orchestra of their thoughts, emotions, and awareness. As the mind's chaotic cacophony begins to harmonize through meditative practices, a metaphorical conductor's wand is poised to create a symphony of coherence. In quantum terms, coherence refers to the alignment of quantum waves, and similarly, meditation aligns the mental and emotional frequencies, fostering a state of profound inner harmony.

Brain Waves and Quantum Resonance: The Synchronized Ballet

Meditation has been associated with shifts in brain wave patterns, with some states aligning with enhanced focus, relaxation, or transcendental experiences. Quantum resonance suggests that the synchronized dance of quantum particles influences the vibrational harmony within the quantum field. This convergence invites contemplation on how the synchronized ballet of brain waves during meditation resonates with the subtle frequencies of the quantum realm.

Just as the act of observation influences quantum outcomes, the focused awareness in meditation may shape the inner quantum landscape. The intentional observation of thoughts and emotions during meditation invites individuals to recognize their influence on the unfolding quantum realities within consciousness. This alignment encourages a deeper understanding of how the meditator becomes an active participant in the co-creation of their inner quantum universe.

Quantum entanglement reveals the interconnected nature of particles, and similarly, meditation often leads to a sense of oneness, a feeling of unity with oneself, others, and the cosmos. This convergence invites individuals to explore the possibility that the unity experienced in meditation resonates with the profound interconnectedness inherent in the quantum fabric of reality.

In the meditative state, focused intentions may influence the quantum field of possibilities, potentially leading to shifts in perception, well-being, or personal growth. This alignment encourages individuals to recognize the co-creative interplay between conscious intention and the quantum potentials that unfold during moments of profound stillness.

As we navigate the intertwining realms of meditation and quantum coherence, this section invites individuals to explore the transformative landscape where the art of meditation becomes a gateway to the quantum dimensions within. It unfolds a harmonious symphony where the intentional alignment of mind and consciousness resonates with the subtle frequencies of the quantum universe, offering a profound journey into the interconnected dance of mind and matter.

Quantum Mindfulness: Integrating Awareness and Being

In the quest for unity, we seek a holistic worldview that harmoniously integrates scientific inquiry and spiritual wisdom. By embracing the interconnected nature of all things, we envision a paradigm where science and spirituality coalesce, offering a comprehensive understanding of reality that transcends disciplinary boundaries. Quantum Mindfulness, a dynamic approach that transcends traditional notions of mindfulness by integrating awareness not only with the mind but with the entire quantum body. In this section, we discuss the profound dimensions of Quantum Mindfulness, where the present moment unfolds as a rich tapestry of quantum processes, inviting a deep connection to the essence of being.

At the core of Quantum Mindfulness lies the practice of Quantum Presence, an invitation to fully embrace the present moment. Rather than solely focusing on mental awareness, Quantum

Presence extends to the quantum body, fostering a profound connection to the intricate movement of quantum processes unfolding within and around us. This expanded awareness opens doors to a deeper understanding of the now as a convergence of quantum possibilities.

Traditional mindfulness often emphasizes mental presence, but Quantum Mindfulness expands this awareness to include the entire quantum body. By tuning into the subtle energies, sensations, and quantum processes within the body, individuals can cultivate a more holistic and embodied form of mindfulness, enriching the depth of conscious experience.

Quantum Breathing is a practice that aligns the rhythm of breath with conscious awareness. In Quantum Mindfulness, each breath becomes a journey into the quantum dimensions of consciousness. The inhale and exhale mirror the ebb and flow of quantum waves, creating a harmonious resonance between the breath, the quantum body, and the present moment. Quantum Breathing serves as a gateway to heightened states of awareness and inner tranquility.

Engage in the practice of Observing Quantum Thoughts, where mindfulness extends beyond the mental realm to include the quantum nature of thoughts. By observing thoughts with detached awareness, individuals can recognize the transient and probabilistic nature of mental processes. This practice invites a shift from identification with thoughts to a state of quantum awareness where the observer and the observed become intimately entwined.

This practice involves becoming attuned to the quantum dance of sensations, recognizing the interconnected web of energies that contribute to the tapestry of conscious experience. Quantum Sensations offer a gateway to a deeper understanding of the quantum self.

In Non-Dual Awareness within Quantum Mindfulness, the illusion of separation dissolves, and unity prevails. By recognizing the interconnected nature of the quantum self with the broader quantum reality, individuals can experience a profound sense of oneness. Non-Dual Awareness invites a shift from dualistic thinking to a holistic perception of the quantum tapestry that unites all aspects of being.

As we immerse ourselves in Quantum Mindfulness, the integration of awareness and being takes on new dimensions, inviting individuals to explore the quantum depths of consciousness. This dynamic approach transcends the boundaries of traditional mindfulness, offering a transformative journey into the interconnected field of the quantum self and the timeless present moment.

Chapter 5: Embracing the Infinite Presence

Practices for staying connected to the present moment:

Rooted in the principles of mindfulness and inspired by the vastness of the quantum cosmos, this chapter guides you on a journey to cultivate an enduring connection with the present moment. From recognizing the boundless nature of awareness to practical exercises that anchor one in the infinite now, each section invites individuals to embrace a timeless and expansive sense of presence.

The Boundless Horizon of Awareness:

In this section, we navigate the expansiveness of consciousness, drawing inspiration from the vastness of the quantum cosmos. The

metaphorical horizon leads us to recognize that awareness extends far beyond the limits of our everyday perceptions, inviting a transformative shift in our understanding of the self and the present moment.

In the exploration of the boundless horizon, we encounter the idea that consciousness knows no boundaries, transcending the constraints of time and space. Through mindful practices, you can gradually dissolve the perceived edges of consciousness, embracing the vista that unfolds within the boundless horizons of the present.

As we venture deeper into this metaphorical terrain, we discover that the boundless horizon of awareness allows us to witness the interconnectedness of all things. In this interconnected awareness, the distinctions between self and other, observer and observed, begin to blur, offering a profound sense of unity with the unfolding quantum reality.

By cultivating an understanding that the horizon of consciousness extends infinitely, you can enhance your ability to stay present in each moment. This practice becomes a gateway to profound mindfulness, where the boundless nature of awareness becomes a guiding light in the journey of self-discovery.

The Boundless Horizon of Awareness persuades us to transcend the ordinary confines of perception, inviting a mindful expansion into the limitless dimensions of consciousness. The horizon unfolds as a canvas of infinite possibilities, encouraging you to embrace the timeless and expansive nature of the present moment.

The Eternal Now: Navigating Timelessness

At the heart of this exploration lies the recognition that each moment contains the timeless essence of the eternal now. You can navigate the intricacies of timelessness within your conscious experience. By embracing the present as an ever-unfolding moment, you can transcend the conventional constraints of past and future, anchoring yourself in the perpetual flow of the eternal now. By practicing mindful awareness, you can attune yourself to the inherent richness and depth present in each moment. This section serves as a compass, guiding you to navigate the currents of the eternal now with intention and presence.

As we venture into the exploration of The Eternal Now, we discover that time, as experienced in the quantum realm, is not a linear progression but rather a dynamic interplay of possibilities. By aligning with this timeless dimension, you can cultivate a heightened sense of awareness and presence. This practice becomes a transformative journey, allowing you to navigate the river of time with grace, appreciating the beauty and significance inherent in every passing moment.

The Eternal Now calls us to relinquish the constraints of past and future, inviting a conscious immersion into the timeless flow of the present. May the eternal now become a guiding beacon, illuminating the path to a more profound and connected experience of the ever-present moment.

Quantum Flow: Riding the Wave of the Present

Just as particles in the quantum realm exhibit a dynamic and fluid nature, so too does the essence of conscious experience. The metaphor of Quantum Flow becomes a guiding principle, encouraging you to perceive each moment as a wave in the vast ocean of the present. By recognizing the ever-changing nature of the present, you can learn to surf the waves of experience, cultivating a harmonious dance with the perpetual flow of quantum reality.

You can learn to adapt and resonate with the fluidity of each moment, enhancing your capacity to navigate life's ever-shifting circumstances. This section serves as a guide, encouraging you to approach the present with a sense of openness, curiosity, and a willingness to ride the waves of experience.

As we reveal Quantum Flow, we discover that the present moment is not static but rather a dynamic and interconnected tapestry of experiences. By embracing the fluidity of the now, individuals can cultivate a deeper sense of presence and resilience, learning to dance with the ebb and flow of life. This practice becomes an invitation to let go of resistance, surrendering to the natural rhythm of the present.

Quantum Flow allows us to release the need for control and embrace the inherent movement within the present moment. May the metaphor of riding the wave inspire a profound connection to the flow of experience, fostering a mindful engagement with the ever-changing and vibrant nature of the quantum reality we inhabit.

Quantum Expansion: Radiating Presence Beyond Limits

At the heart of Quantum Expansion lies the recognition that our awareness has limitless horizons. Like particles occupying vast states in the quantum realm, you can consciously expand your consciousness beyond conventional boundaries. This section encourages you to embrace the boundless potential within, cultivating a sense of presence that transcends self-imposed limits. By radiating awareness beyond perceived limitations, readers can discover the richness and expansiveness of their own conscious existence.

We discover that personal growth is not confined to linear progression but rather an ongoing, dynamic process. By aligning with the expansive nature of the quantum cosmos, you can foster a deeper connection to your own potential. This practice becomes an invitation to step beyond the familiar, embracing the uncharted territories of personal evolution.

Quantum Expansion allows us to release the constraints of self-imposed boundaries, inviting a conscious and intentional journey toward personal growth; fostering a mindful expansion that mirrors the boundless nature of the quantum reality we inhabit.

The Quantum Pause: Finding Stillness in Motion

In the dance of particles within the quantum landscape, there exists a fascinating phenomenon, the Quantum Pause. It is a moment of apparent stillness within the constant motion, reminiscent of the quiet interludes between musical notes. This section encourages you to recognize and embrace your own Quantum Pause, a mindful technique for finding serenity in the midst of life's bustling activity.

We discover that stillness is not the absence of motion but a harmonious presence within it. By consciously integrating moments of pause, you can cultivate a deeper connection to the present moment. This practice becomes an invitation to appreciate the subtle beauty found in the pauses between thoughts, actions, and experiences. The Quantum Pause will show us the profound serenity that arises when we intentionally find stillness in motion.

Here are practical exercises tailored to cultivate awareness, presence, flow, expansion, and stillness:

1. The Boundless Horizon of Awareness: Expanding Consciousness

Exercise: Find a quiet space and sit comfortably. Close your eyes and take a few deep breaths. Visualize your awareness expanding beyond the boundaries of your body. Envision it reaching out like ripples on water, extending to the farthest reaches of your imagination. Hold this visualization, acknowledging the boundless nature of your awareness. Take note of any shifts in perception or sensations. Repeat regularly to enhance your sense of expanded consciousness.

2. The Eternal Now: Timeless Awareness Practice

Exercise: Choose a simple activity you do daily, such as washing dishes or walking. Engage in the activity with full attention, focusing on each moment without thinking about what happened before or what might come after. Allow yourself to become fully immersed in the present task. Notice the sensory details; the feeling of water, the texture of surfaces, or the rhythm of your steps. Whenever your mind drifts to the past or future, gently bring it back to the current moment. Practice this regularly to strengthen your connection to the eternal now.

3. Quantum Flow: Mindful Adaptability

Exercise: Select a dynamic activity like dancing or flowing yoga sequences. As you engage in the movement, pay close attention to the transitions between poses or steps. Embrace the fluidity of the motion, allowing your body to adapt naturally. Be present with the sensations, breath, and rhythm. If your mind resists the flow, consciously bring your awareness back to the movement. This exercise fosters mindful adaptability and a sense of ease in navigating life's changes.

4. Quantum Expansion: Breaking Personal Boundaries

Exercise: Identify a belief or habit that feels limiting or constrictive. Write it down and explore why you hold onto it. Then, envision an expanded version of yourself without this limitation. Imagine the possibilities, skills, or qualities you would embody. Consciously choose to step into this expanded version in small, manageable steps. Note any shifts in your perception or experiences. Repeat the process with different aspects of your life regularly to foster a mindset of continuous growth.

5. The Quantum Pause: Mindful Stillness Amidst Action

Exercise: Integrate intentional pauses into your daily routine. For instance, pause for a moment before responding to emails, taking a deep breath to center yourself. While commuting, practice a brief moment of stillness before starting your vehicle or stepping onto public transportation. Use these pauses to reconnect with your breath and bring awareness to the present moment. Over time, these intentional breaks will become opportunities for rejuvenation and mindful awareness amidst the busyness of life.

These practical exercises aim to assist you in cultivating mindfulness, expanding your consciousness, embracing the present moment, adapting to life's flow, breaking personal boundaries, and finding stillness within motion. They can be adapted to suit various lifestyles and preferences, offering accessible tools for personal growth and well-being.

Chapter 6: The Quantum Mind Unveiled

Exploring the Mysteries of the Mind from a Quantum Perspective:
Here, the intersection of quantum principles and the intricacies of consciousness becomes a captivating exploration. As we dig into the mysteries of the mind, we unveil the untapped potential residing within the quantum fabric of our cognitive existence.

Decoding Quantum Algorithms: The Mind as a Quantum Processor
Imagine the mind as a sophisticated quantum processor, capable of processing information in ways that go beyond the limitations of classical computation. This metaphorical decoding of Quantum Algorithms sheds light on the parallelism and superposition within cognitive processes. It is essential to recognize the

quantum nature of your own thought patterns, fostering a deeper understanding of the mind's capacity for simultaneous, multidimensional processing.

Superposition of Ideas: Embracing Multifaceted Thinking

In the concept of Superposition of Ideas, thoughts exist in a state of potentiality, simultaneously occupying multiple perspectives. The mind can entertain many diverse possibilities at once, fostering creativity and innovative thinking.

Quantum Coding of Intuition: Tapping into Inner Wisdom

In the quantum coding of intuition, the mind accesses a reservoir of inner wisdom beyond logical reasoning. Intuition, like a quantum algorithm, operates in a nonlinear fashion, providing insights that transcend conventional thought processes. Practical exercises guide readers to tap into their intuitive capacities, recognizing and decoding the subtle messages that arise from the quantum depths of the mind.

Quantum Uncertainty: Embracing the Unknown

When we look at the role of Quantum Uncertainty in cognitive processes, we recognize that the mind's potential extends into realms of the unknown. Uncertainty becomes a catalyst for creativity, innovation, and the emergence of novel ideas. Recognizing the mind's capacity for simultaneous consideration of diverse viewpoints is a powerful tool for expanding creative thinking. This approach nurtures a mindset that revels in the interconnectedness of thoughts, pushing the boundaries of conventional thinking.

Practical Application: Quantum Journaling

Initiate a Quantum Journaling practice. Set aside dedicated time to jot down thoughts, ideas, or concepts without the pressure of coherence or linearity. Instead of organizing them in a structured manner, allow them to exist on the page in a superposition of ideas. Upon reflection, observe the relationships and unexpected connections that emerge. This exercise promotes a flexible and open-minded approach to thinking, unlocking the creative potential inherent in the superposition of ideas.

As we decipher The Quantum Code, this section serves as a bridge between quantum principles and the cognitive mysteries within the mind. Through metaphors and practical applications, readers are invited to explore the profound implications of the mind as a quantum processor, unlocking the potential for enhanced cognitive understanding and the unraveling of mysteries that lie within the quantum fabric of conscious thought.

Quantum Cognition: Navigating Nonlinear Thought

The mind's nonlinear operation is a captivating choreography that nurtures creativity, intuition, and the birthing of innovative ideas. In the fluidity of this dance, traditional boundaries and linear constraints are cast aside, allowing for a dynamic interplay of thoughts and emotions.

In the nonlinear motion of the mind, creativity is set free from the shackles of predictable patterns. Ideas converge and diverge, forming unexpected connections and sparking novel insights. The mind, free from the constraints of linear thinking, becomes a boundless canvas where creative expression flourishes. This dance enables individuals to explore unconventional perspectives, encouraging the synthesis of diverse elements into unique and imaginative concepts.

Intuition, akin to a partner in this dance, emerges as a guiding force. Freed from the constraints of logic, the nonlinear mind intuitively navigates the complex web of thoughts, drawing upon subtle cues and deeper insights. In the dance's ebb and flow, intuition becomes a powerful ally, guiding individuals toward innovative solutions and fostering a profound connection to their inner wisdom.

Novel ideas often arise from the spontaneous movements of the nonlinear mind. As thoughts intertwine and leap beyond the conventional boundaries, the dance becomes a fertile ground for the emergence of groundbreaking concepts. The nonlinear mind embraces the unexpected, giving rise to ideas that challenge the status quo and propel individuals into unexplored intellectual territories.

Practical Application: Mindful Creativity Sessions

Engage in mindful creativity sessions to nurture the nonlinear functioning of the mind. Set aside dedicated time for activities such as free-form writing, drawing, or brainstorming without predetermined structures. Allow thoughts to flow without judgment or the need for immediate coherence. Embrace the unexpected connections that arise during these sessions, acknowledging that creativity flourishes in the spontaneity of the nonlinear dance.

As we explore the mind's nonlinear dance, may we recognize its role as a catalyst for creativity, intuition, and the emergence of novel ideas. This dance is an invitation to embrace the fluidity of thought, encouraging us to move beyond the confines of linear thinking and step into the uncharted realms where innovation thrives.

Quantum Memory: Subjective Recall of the Observer

Memory, far from being a linear playback, unveils itself as a dynamic and intricate process of reconstruction intricately influenced by the observer. The mind engages in a continual process of reshaping and reinterpreting past experiences, forging a mosaic of memories that is both malleable and subjective.

Nonlinear Memory Reconstruction:
Memory is not akin to a static film reel playing back events in a linear sequence. Instead, it is a nonlinear reconstruction, where fragments of experiences are woven together in a fluid, ever-evolving narrative. The mind, acting as both curator and storyteller, rearranges these fragments, emphasizing certain details while downplaying others, based on the observer's perspective and emotional state.

Influence of Perception and Emotion:
The observer plays a crucial role in shaping the narrative. Perception and emotion become integral partners, influencing the selection and interpretation of memories. What stands out vividly or fades into the background is subject to the observer's current mindset, biases, and emotional associations. Thus, the same memory

can be reconstructed in diverse ways, depending on the observer's frame of mind.

Constructive Nature of Memory:

Memory is inherently constructive, akin to an ongoing work of art that evolves with each recollection. As memories are retrieved, they undergo subtle alterations influenced by the present moment. New information, emotions, or perspectives acquired since the initial experience can blend seamlessly into the memory, contributing to its ongoing reconstruction.

Practical Application: Mindful Reflection Exercises

Engage in mindful reflection exercises to explore the dynamic nature of memory. Select a significant past event and intentionally revisit it, paying attention to the emotions and thoughts that arise. Notice how the memory evolves with each recollection, and be mindful of the influence of your current perspective on the reconstruction. This practice fosters awareness of the malleability of memory and promotes a deeper understanding of the constructive nature of recollection.

As we unravel the complexities of memory, let us appreciate its nonlinear, dynamic nature, the past and present influenced by the observer's lens. Understanding memory as a continuously evolving narrative enriches our appreciation for the subjective and intricate process of recollection.

Conscious Quantum Evolution: Shaping the Mind's Potential

Let's explore the transformative concept of Conscious Quantum Evolution, where intentional awareness emerges as a powerful catalyst for shaping the potential of the mind. In this exploration, mindfulness practices intertwine with quantum principles, offering a pathway to influence cognitive evolution in profound ways.

Intentional Awareness as Catalyst:
At the heart of Conscious Quantum Evolution is the recognition that intentional awareness acts as a catalyst for change. By directing conscious attention and intention towards specific aspects of cognitive growth, individuals become active participants in the evolution of their own minds. This intentional focus serves as a guiding force, shaping the trajectory of cognitive development in alignment with personal and collective aspirations.

Mindfulness Aligned with Quantum Principles:
Mindfulness practices, when harmonized with quantum principles, form a symbiotic relationship that amplifies their impact. Quantum principles emphasize the interconnectedness of all things and the observer's role in shaping reality. Mindfulness, rooted in the present moment, aligns with these

principles by fostering a heightened state of awareness and encouraging individuals to observe their thoughts without attachment or judgment.

Quantum Entanglement of Consciousness:
The concept of Quantum Entanglement of Consciousness takes center stage. As intentional awareness expands, it becomes entangled with the quantum fabric of reality. This interconnectedness implies that personal cognitive evolution is not isolated but influences and is influenced by the broader consciousness of the collective. Mindfulness practices, by enhancing this entanglement, contribute to a shared evolution that extends beyond individual boundaries.

Practical Application: Quantum Mindfulness Meditation

Set aside dedicated time for meditation, focusing on intentional awareness and the alignment of mindfulness with quantum principles. During this practice, visualize the interconnectedness of your consciousness with the quantum field. Direct your awareness toward aspects of cognitive growth you wish to nurture. This intentional alignment amplifies the potency of mindfulness, fostering a conscious evolution of the mind.

As we explore Conscious Quantum Evolution, let us embrace the idea that intentional awareness is not merely an observer of cognitive processes but an active agent in their evolution. The intertwining of mindfulness with quantum principles becomes a transformative journey, inviting individuals to shape the potential of their minds consciously and contribute to the collective evolution of consciousness.

The quantum principles intertwined with the cognitive landscape unveils not only the complexity but also the extraordinary potential within the human mind. Through practical insights and exercises, individuals are empowered to navigate the quantum terrain of their own consciousness, unlocking new dimensions of understanding, creativity, and cognitive evolution.

Chapter 7: Harmony of Thought:

Integrating Mindfulness into Daily Life:
Harmony of Thought emerges as a melodic
composition where thoughts and emotions
move in synchrony. This chapter invites you to
weave the threads of mindfulness into the
tapestry of daily life to cultivate a harmonious
and balanced existence.

The Dance of Thoughts and Emotions:
Within the vastness of consciousness there is a
harmonious or discordant interplay between
your thoughts and emotions. This dance is a
profound manifestation of their
interconnectedness, where the quality of
thoughts serves as a subtle maestro, shaping the
landscapes of our emotional experiences.

The Interwoven Tapestry:

Thoughts and emotions are threads in the intricate tapestry of consciousness, weaving together to create the fabric of our subjective experience. In the dynamic relationship between them, we can see that each thought contributes to the emotional landscape and every emotion colors the thoughts that follow. This symbiotic relationship reveals the inseparable nature of thoughts and emotions.

Influence of Thought Quality on Emotions: Positive, constructive thoughts have the potential to cultivate joy, gratitude, and optimism, creating a harmonious emotional ambiance. Conversely, negative or self-critical thoughts may give rise to emotions such as anxiety, sadness, or frustration. Through mindful observation, you gain insight into how the nature of your thoughts influences the emotional hues of your internal landscape.

The Feedback Loop:

Recognize the feedback loop between thoughts and emotions. Thought patterns can set the stage for emotional responses, and in turn, emotions can shape the nature of subsequent thoughts. This cyclic movement is both subtle and powerful, impacting our overall well-being. By understanding this feedback loop, you gain the agency to intervene mindfully, disrupting patterns that may contribute to emotional discord.

Mindful Awareness:
Become a mindful observer of this interplay, cultivating awareness of the thoughts that arise and the emotions they evoke. In moments of heightened emotion, practice stepping back and examining the accompanying thoughts. Similarly, when engaged in thought patterns, observe how they influence your emotional state. This intentional awareness forms the foundation for fostering a more harmonious relationship between thoughts and emotions.

Practical Application: Mindful Thought Check-In:

Engage in a Mindful Thought Check-In throughout the day. Set aside brief moments to observe the prevailing thoughts and accompanying emotions. Notice patterns and connections between specific thoughts and emotional responses. This practice fosters a conscious understanding of the dance between thoughts and emotions, empowering you to navigate this interplay with greater clarity.

As you navigate the intricate dance between thoughts and emotions, may this exploration serve as a guide to recognizing their profound interconnectedness. Through mindful awareness, individuals can consciously shape the quality of their thoughts, influencing the emotional landscapes that unfold within, and fostering a harmonious dance within the symphony of consciousness.

The Mindful Observer: Cultivating Awareness:

Intentional awareness emerges as a potent force, an observer that witnesses the ebb and flow of mental and emotional states. This mindful presence holds the transformative power to navigate the complex landscapes of consciousness with clarity and equanimity.

The Mind as an Ever-Changing Landscape: Visualize the mind as a vast and ever-changing landscape, with thoughts and emotions as the natural elements that ebb and flow. Intentional awareness is the panoramic view, allowing you to witness the subtle nuances of this internal terrain. Embrace the understanding that, like nature, the mind has its seasons, sometimes calm, sometimes stormy, and intentional awareness provides the lens through which to observe these fluctuations.

Non-Judgmental Observation: The power of intentional awareness lies in its non-judgmental quality. As the mindful observer, resist the urge to label thoughts and emotions as inherently good or bad. Instead, observe them with a sense of curiosity and acceptance. This impartial observation serves as a bridge to understanding, allowing you to witness the complexities of your mental and emotional states without being swept away by their currents.

Detachment and Equanimity:
Intentional awareness fosters a sense of detachment, a conscious stepping back from the immediate identification with thoughts and emotions. This detachment does not diminish the richness of experience; rather, it provides a space for equanimity. By cultivating equanimity, you become less entangled in the intensity of mental and emotional fluctuations, allowing for a more balanced and grounded experience.

Riding the Waves of Experience:
The ebb and flow of mental and emotional states become a dynamic dance, and intentional awareness equips you to ride the waves of

experience with grace. Whether in moments of joy, sorrow, excitement, or calm, the mindful observer remains steadfast, providing a stable vantage point amidst the changing tides. This capacity to ride the waves enhances resilience and promotes a deeper connection to the essence of each moment.

Practical Application: Mindful Breathing Anchoring:

Utilize the practice of mindful breathing as an anchoring technique. During moments of heightened mental or emotional activity, pause and turn your attention to the breath. Observe each inhalation and exhalation intentionally, allowing the breath to serve as an anchor in the present moment. This simple yet powerful practice strengthens the connection to intentional awareness, offering a stable point amidst the fluctuating landscape of the mind.

As you uncover the power of intentional awareness, may it become a guiding light, illuminating the ebb and flow of your mental and emotional states. Through this intentional observation, individuals gain the ability to navigate the ever-changing landscape of consciousness with a mindful presence that fosters clarity, resilience, and a deep appreciation for the richness of experience.

Embracing Emotional Intelligence:
A profound connection emerges between mindfulness and emotional intelligence. The practice of mindfulness becomes a transformative journey, cultivating a heightened sensitivity and responsive awareness to the intricate language of emotions.

Mindfulness as the Gateway to Emotional Intelligence:
Mindfulness serves as a gateway to unlocking the depths of emotional intelligence. By directing intentional awareness to the present

moment, you create a space for understanding and embracing the nuances of your emotional landscape. This practice becomes the foundation for developing emotional intelligence a skill that involves recognizing, understanding, and managing one's own emotions and empathizing with the emotions of others.

Heightened Sensitivity to Emotional Cues: Through mindfulness, sensitivity to emotional cues becomes refined. The mindful practitioner develops an acute awareness of subtle shifts in emotional states; whether they manifest as a fleeting thought, a subtle bodily sensation, or a nuanced change in mood. This heightened sensitivity allows for early recognition of emotions, fostering a proactive approach to emotional well-being.

Cultivating a Non-Reactive State: Mindfulness encourages a non-reactive awareness of emotions. Rather than impulsively responding to emotional stimuli, you learn to observe emotions with equanimity. This non-reactive stance provides a valuable pause, allowing for a thoughtful and intentional response to emotional experiences. This mindful approach becomes the cornerstone of emotional intelligence, empowering you to navigate the complexities of your internal world with grace.

Empathy and Compassion as Natural Outcomes: Mindfulness nurtures the seeds of empathy and compassion. As you become attuned to your own emotions, you naturally develop a deeper understanding of the emotions of others. This empathetic connection is grounded in the shared human experience of joy, sorrow, and everything in between. Mindfulness fosters a compassionate responsiveness, encouraging you to connect with others on a profound emotional level.

Practical Application: Mindful Emotional Check-Ins:

Incorporate Mindful Emotional Check-Ins into your daily routine. Set aside moments for intentional self-reflection, observing and labeling your current emotional state without judgment. Notice the sensations associated with each emotion and explore the thoughts that accompany them. This practice enhances emotional self-awareness, laying the groundwork for the development of emotional intelligence.

As you embrace the profound connection between mindfulness and emotional intelligence, may this exploration serve as a guide to cultivating a heightened sensitivity and responsiveness to the intricate tapestry of emotions. Through the practice of mindfulness, individuals embark on a transformative journey toward emotional intelligence—a journey that deepens self-awareness, nurtures empathy, and fosters a compassionate engagement with the ever-changing landscape of feelings.

Mindful Decision-Making: The Path to Clarity
In the labyrinth of decision-making, intentional awareness emerges as a guiding light, illuminating the path to clarity and discernment. The practice of mindfulness becomes a transformative tool, empowering you to make choices aligned with their values and aspirations.

Illuminating Values and Aspirations:
Mindful awareness acts as a spotlight, directing attention to the core values and aspirations that shapes your journey. Through intentional reflection, you gain a profound understanding of what truly matters to you. This illuminating process clarifies the lens through which decisions are made, ensuring that choices resonate with the deeper currents of personal meaning.

Creating Space for Reflection:
Mindfulness creates a spacious arena for reflective thought. In the midst of life's busyness, intentional awareness allows you to step back

and examine your thoughts and feelings about a decision. This space for reflection is essential for discerning the nuances of various options and understanding the potential impact of each choice on your life path.

Embracing Present-Moment Clarity:
The present moment becomes a crucible of clarity. Mindful attention to the immediate experience provides a clear perspective unburdened by past regrets or future anxieties. By fully engaging with the present, you can make decisions anchored in the clarity of the current reality, unclouded by the noise of distractions or irrelevant concerns.

Discernment Amidst Complexity:
Intentional awareness enhances discernment in the face of complexity. Mindfulness encourages you to break down complex decisions into manageable components, examining each aspect with a focused and open mind. This discerning approach enables a more nuanced

understanding of the potential outcomes, facilitating choices that align with overarching values and aspirations.

Aligned Action and Authenticity:
Mindful decision-making culminates in aligned action. Choices made with intentional awareness are inherently authentic, reflecting the true essence of an your values and aspirations. This authenticity fosters a sense of congruence, where decisions resonate with the inner core, leading to a more fulfilling and purposeful life path.

Practical Application: Mindful Decision Journaling:

Create a Mindful Decision Journal as a practical tool for intentional awareness in decision-making. Document key decisions, the thought processes involved, and the emotions experienced. Reflect on how each decision aligns with your values and aspirations. This practice enhances the clarity and discernment applied to future choices, creating a record of mindful decision-making.

As you traverse the terrain of decision-making, may intentional awareness guide you toward clarity and discernment. Through mindfulness, individuals can navigate the complexities of choices with a deep understanding of their values, aspirations, and the present moment, forging a path that authentically aligns with the essence of who they are and wish to become. Through intentional awareness and mindful integration, you are initiating a transformative path toward a balanced and enriched way of being.

Chapter 8: Transforming Your Life

Real-life Examples of Applying Quantum Mindfulness Exercises and Practices for Immediate Implementation:

In this transformative chapter, we detail the practical applications of Quantum Mindfulness, providing real-life examples that illustrate its profound impact on everyday life. Through a series of exercises and practices, you'll discover how to integrate Quantum Mindfulness into your reality, fostering immediate transformation and enriching your journey towards self-mastery. Here are the practical exercises.

Quantum Breath Integration:

Quantum Breath Integration is a transformative practice that marries intentional breathwork with the principles of quantum consciousness. This immersive experience invites you to explore the dimensions of your own consciousness with each breath, forging a profound connection between your breath and the quantum fabric of reality.

1. Setting the Stage: Mindful Preparation

Begin by finding a quiet and comfortable space where you won't be disturbed. Assume a relaxed posture, whether sitting or lying down, and bring your attention to the present moment. Ground yourself in the awareness of your body and the immediate surroundings.

2. Intentional Awareness: Aligning with Quantum Principles

Start by setting an intention for your breathwork. This could be to deepen your connection with the quantum dimensions of your consciousness, cultivate inner harmony, or explore a specific aspect of self-awareness. The key is to anchor your breath with purpose, aligning it with the quantum principles of interconnectedness and potentiality.

3. Mindful Inhalation: Inhaling the Quantum Essence

With each inhalation, visualize the breath as a conduit for the quantum essence that permeates your being. Imagine drawing in not just air but the very energy that connects all things in the quantum realm. Feel the expansiveness of each breath, inviting the quantum potential into every cell of your body.

4. Quantum Resonance: Experiencing the Vibrational Harmony

As you continue to breathe intentionally, tune into the vibrational harmony within. Envision your breath as a resonance that aligns with the quantum frequencies of your consciousness. Feel the subtle energies within and around you, creating a symphony of coherence between your breath and the quantum field.

5. Exhalation as Release: Letting Go into Quantum Possibilities

During the exhalation, release any tension, thoughts, or emotions that no longer serve you. Envision exhaling into the quantum field, surrendering to the infinite possibilities that reside there. Each exhale becomes a conscious act of letting go, clearing the path for new quantum potentials to unfold.

6. Quantum Breath as a Journey: Breath by Breath Exploration

View each breath as a journey into the quantum dimensions of your own consciousness. With each cycle, explore different facets of your inner landscape. Notice how the quality of your breath reflects and influences your state of mind, creating a dynamic interplay between breath and consciousness.

7. Closing the Practice: Integration and Reflection

As you conclude the Quantum Breath Integration, take a few moments to integrate the experience. Notice any shifts in your awareness, emotions, or overall sense of well-being. Consider jotting down your reflections in a journal, capturing the nuances of your journey into quantum consciousness through intentional breathwork.

8. Regular Practice: Weaving Quantum Breath into Daily Life

Make Quantum Breath Integration a regular practice, seamlessly weaving it into your daily routine. Whether as a morning ritual, a tool for stress relief during the day, or a mindful transition before sleep, regularity enhances the transformative power of this practice.

Through intentional breathwork, you unlock the door to a deeper understanding of self, fostering a harmonious connection with the infinite possibilities that reside within the quantum fabric of reality.

Mindful Quantum Leap Journal: A Comprehensive Guide

The Mindful Quantum Leap Journal is a dynamic tool designed to track your conscious shifts in perception and explore the transformative journey of Quantum Mindfulness. This journal serves as a personal space for reflection, capturing moments of intentional awareness and providing a roadmap for your evolution. Let's delve into the details of how to effectively use and derive maximum benefit from this journal.

1. Setting Up Your Journal:
Begin by dedicating a specific notebook or digital document for your Mindful Quantum Leap Journal. Choose a format that resonates with you, whether it's a physical journal, a digital app, or a combination of both. The goal is to create a sacred space where you can chronicle your mindful journey.

2. Daily Entries: Recording Quantum Shifts
Commit to making daily entries in your journal. Start by noting the date and time of your entry. Reflect on moments of intentional awareness throughout the day; times when you consciously shifted your perception, engaged in mindful practices, or observed the world with a heightened sense of presence.

3. Capturing Thoughts and Emotions:
Record your thoughts and emotions during these mindful moments. Be specific and descriptive. What triggered your intentional awareness? How did you feel? What thoughts arose during these shifts? This detailed documentation allows you to trace patterns, identify growth areas, and gain insights into the nature of your consciousness.

4. Visualization and Symbolism:
Enhance your journal entries with visualizations and symbols. If a particular moment of mindful awareness was accompanied by a vivid mental

image or symbol, sketch or describe it in your journal. Visual representations can deepen your connection to the experience and serve as powerful anchors for future reflections.

5. Intention Setting:

Use your journal as a space for setting intentions. Outline specific areas of your life or aspects of consciousness that you wish to explore or transform. Setting clear intentions provides direction for your mindful journey and becomes a guide for intentional awareness in daily life.

6. Weekly Reflections:

Dedicate a section of your journal for weekly reflections. Review your daily entries, looking for patterns, recurring themes, or shifts in your mindful experiences. Reflect on the impact of intentional awareness on your overall well-being and personal growth. Consider how these insights align with your broader goals and aspirations.

7. Gratitude and Acknowledgment:
Incorporate gratitude into your journal practice. Express gratitude for moments of mindfulness, personal growth, and the discoveries you make along the way. Acknowledge the effort you put into intentional awareness and celebrate the small victories in your journey toward self-mastery.

8. Challenges and Learnings:
Document challenges and learnings. If you encounter obstacles in maintaining intentional awareness or experience moments of resistance, record them in your journal. Explore potential insights or lessons embedded in these challenges, turning them into opportunities for growth.

9. Goal Setting and Evolution:
Periodically revisit your initial intentions and set new goals for your Quantum Mindfulness practice. Use your journal as a tool for tracking your evolution, celebrating milestones, and

refining your path forward. Consider integrating new mindful practices or adjusting existing ones based on your experiences.

10. Reflecting on Quantum Leaps:
Celebrate moments of significant shifts in perception—your "Quantum Leaps." When you look back over the weeks or months, identify those instances where intentional awareness led to transformative insights or changes in your perspective. Reflect on the impact of these Quantum Leaps on your overall well-being and personal development.

11. Personalizing Your Journal:
Feel free to personalize your journal to make it uniquely yours. Add quotes, affirmations, or images that inspire you. Experiment with different formats, colors, or creative elements that resonate with your individual style. Your Mindful Quantum Leap Journal is a canvas for self-expression and exploration.

12. Reviewing and Revisiting:
Regularly review and revisit your journal entries. This retrospective practice allows you to witness your journey unfold, observe patterns, and appreciate the continuous evolution of your consciousness. It becomes a valuable source of inspiration and a testament to the transformative power of intentional awareness.

As you engage with your Mindful Quantum Leap Journal, remember that it's not just a record of your experiences but a living document that catalyzes growth. Use it as a compass for your mindful journey, guiding you through the dimensions of Quantum Mindfulness and illuminating the path to self-discovery and self-mastery.

Quantum Observation Walk:

The Quantum Observation Walk is a practice that transforms your daily walk into a mindful exploration of the present moment. This immersive experience integrates intentional awareness with the act of walking, inviting you to observe and connect with the world around you in a quantum-conscious way. Let's explore the intricacies of this practice to make the most of your Quantum Observation Walk.

1. Preparation: Setting the Mindful Tone
Before you start your Quantum Observation Walk, take a moment to set the tone for intentional awareness. Find a quiet space to stand still, close your eyes, and take a few mindful breaths. Center yourself in the present moment, leaving behind any distractions or preoccupations.

2. Mindful Stepping: Walking with Intention
As you begin walking, bring your attention to the act of stepping. Feel the ground beneath your feet, the subtle shifts in weight, and the rhythmic pattern of your steps. Walking with intention becomes a meditation in motion,

grounding you in the present and aligning your consciousness with each step.

3. Sensory Awareness: Engaging the Senses

Engage your senses fully. Notice the sounds around you; the rustle of leaves, the chirping of birds, or distant sounds of traffic. Observe the colors, shapes, and textures in your environment. Feel the temperature and breeze on your skin. Let each sensation become an anchor for your awareness, immersing you in the rich tapestry of the present moment.

4. Quantum Observation: Beyond Surface Perception

Shift your observation to a quantum perspective. Instead of merely observing the surface details, delve into the interconnectedness of all things. Recognize that each element; the trees, the sky, even your own breath, is part of a vast quantum field. Observe the subtle energies that connect everything, fostering a sense of unity and oneness.

5. Mindful Breathing Integration: Harmonizing with the Walk

Integrate mindful breathing into your walk. Sync your breath with your steps, creating a harmonious rhythm. With each inhalation, draw in the vitality of the present moment. With each exhalation, release any tension or distractions. This mindful breathing integration enhances your connection with the quantum dimensions of your consciousness.

6. Curiosity and Wonder: Adopting a Childlike Perspective

Approach your walk with a sense of curiosity and wonder, adopting a childlike perspective. Explore your surroundings as if seeing them for the first time. Be open to discovering the beauty in the ordinary, allowing the quantum richness of each moment to unfold.

7. Quantum Timelessness: Transcending Linear Time

Release the constraints of linear time during your Quantum Observation Walk. Allow each step to exist in its own moment, free from past or future concerns. Embrace the timeless nature of the present, where every step becomes a quantum event, unfolding in the eternal now.

8. Gratitude and Acknowledgment: Appreciating the Walk

Express gratitude for the opportunity to engage in this Quantum Observation Walk. Acknowledge the interconnectedness of your own consciousness with the world around you. Feel a sense of appreciation for the quantum dance of existence unfolding with each step.

9. Reflective Pause: Concluding Your Walk Mindfully

As you conclude your Quantum Observation Walk, find a peaceful spot to pause and reflect. Take a moment to revisit the observations and sensations you experienced. Notice any shifts in your awareness or perspective. Carry the quantum essence of your walk with you as you transition back into your daily activities.

10. Regular Practice: Making Quantum Observation a Habit

Make the Quantum Observation Walk a regular practice. Whether it's a daily ritual or a weekly excursion, consistent engagement enhances the transformative power of this mindful walk. Consider exploring different environments, each time approaching the walk with fresh eyes and an open heart.

11. Journaling Your Quantum Walk Experiences:

Complement your Quantum Observation Walk with journaling. Capture your observations, insights, and any shifts in consciousness. Reflect on how this practice influences your overall well-being and perception of the world. Your journal becomes a record of your quantum-conscious journey through mindful walks.

As you embrace the Quantum Observation Walk, may each step be a conscious movement with the quantum dimensions of your consciousness. Through intentional awareness, sensory engagement, and a childlike curiosity, this practice becomes a doorway to the profound interconnectedness of existence, unfolding with each mindful step.

Quantum Living: Integrating Mindfulness into Daily Rituals

Quantum Living is the art of infusing mindfulness into your daily rituals, transforming ordinary moments into opportunities for intentional awareness and self-discovery. This section explores various aspects of your daily life, providing insights and practical suggestions on how to integrate Quantum Mindfulness seamlessly into your routines.

Quantum Morning Routine: Awakening with Intention

Embrace the Quantum Morning Ritual as a sacred gateway to your day. Each element of your morning routine becomes a conscious act, setting the tone for mindful living. Consider the following practices:

1. Mindful Wake-Up Call:

Awaken with gratitude and a mindful breath. Before reaching for your phone or getting out of bed, take a moment to appreciate the gift of a new day. Inhale deeply, exhaling any residual sleepiness, and embrace the present moment.

2. Quantum Shower Meditation:

Turn your morning shower into a Quantum Meditation. Feel the water cascading over your body, focusing on each sensation. Visualize the water cleansing not just your physical body but also your energy field, preparing you for the day ahead.

3. Mindful Breakfast Connection:

Transform your breakfast into a mindful experience. Engage all your senses as you prepare and savor each bite. Consciously appreciate the nourishment your food provides, fostering a deep connection with the energy it imparts.

4. Intention Setting:
Before stepping into the demands of the day, set mindful intentions. Reflect on your goals, aspirations, and the kind of energy you want to bring into your day. Infuse your morning with purpose and clarity, aligning your actions with your broader vision.

Midday Mindfulness: Elevating Energy

Integrate Quantum Midday Mindfulness into your routine to rejuvenate and elevate your energy levels. These practices are designed to create mindful pauses during the day, fostering a quantum-conscious renewal:

1. Mindful Breathing Breaks:
Schedule brief mindful breathing breaks throughout your day. Set an alarm or use natural cues, such as the completion of a task. Take a few minutes to focus on your breath, inhaling deeply and exhaling slowly. This practice renews your energy and enhances your mental clarity.

2. Quantum Lunchtime:
Transform your lunch break into a Quantum Lunchtime. Engage in mindful eating by savoring each bite and appreciating the flavors, textures, and nourishment of your food. Use this time to disconnect from work-related stimuli and immerse yourself in the present moment.

3. Quantum Stretch and Movement:
Incorporate short movement breaks into your midday routine. Stretch your body, stand up, or take a brief walk. Combine movement with mindful breathing to enhance your physical well-being and bring a sense of awareness to the body's sensations.

4. Quantum Focus Reset:

When shifting between tasks, embrace a Quantum Focus Reset. Close your eyes for a moment, take a few intentional breaths, and visualize a clean slate. Release any residual energy from the previous task, and approach the next one with renewed focus and clarity.

5. Gratitude Check-In:

Pause midday for a Gratitude Check-In. Take a moment to reflect on aspects of your day for which you are grateful. This practice shifts your focus toward positive elements, fostering a quantum-conscious perspective and infusing your afternoon with a sense of gratitude.

6. Quantum Visualization Break:

Engage in a Quantum Visualization Break to reset your mental state. Close your eyes and visualize a serene and peaceful scene. Whether it's a natural landscape or a calming image, this practice allows you to recharge and bring a quantum-conscious awareness to your tasks.

Incorporating Quantum Midday Mindfulness into your daily routine creates intentional pauses that elevate your energy, enhance focus, and infuse your workday with mindful consciousness. These brief yet impactful rituals contribute to a quantum-conscious approach to life, fostering well-being and presence in every moment.

Quantum Evening Reflection: Cultivating Closure

End your day with a Quantum Evening Reflection, bringing mindful closure to your daily activities and preparing for restful sleep. Incorporate the following practices:

1. Gratitude Journaling:
Reflect on the day's events in a Gratitude Journal. Write down moments of gratitude, lessons learned, and experiences that brought

joy. This reflective practice fosters a positive mindset and a sense of fulfillment.

2. Quantum Breath Before Sleep:
As you prepare for sleep, engage in a Quantum Breath practice. Inhale deeply, envisioning the breath as a conduit for peace and relaxation. Exhale, releasing any residual tension. Allow your breath to guide you into a restful state, embracing the quantum dimensions of your dreams.

3. Conscious Device Disconnect:
Disconnect from electronic devices mindfully before sleep. Create a buffer zone without screens to allow your mind to transition into a more relaxed state. This conscious disconnection enhances the quality of your rest and supports a quantum-conscious rejuvenation during the night.

4. Quantum Dream Journaling:

Keep a Quantum Dream Journal by your bedside. Upon waking, record any dreams or insights that arise during the night. Dreams often hold symbolic messages from the subconscious, offering a unique window into your inner dimensions.

Embracing Quantum Living is an ongoing journey that transforms your daily rituals into intentional acts of mindfulness. Whether in the morning, within your relationships, or as you wind down in the evening, these practices infuse your life with a quantum consciousness that transcends the ordinary, fostering a rich tapestry of mindful living.

Quantum Presence in Relationships: Cultivating Connection

Integrating Quantum Mindfulness into your relationships fosters deeper connections and enriches shared consciousness. Explore the following practices:

1. Mindful Listening:

In conversations, practice mindful listening. Focus fully on the speaker, setting aside distractions. Observe not just words but also tone, body language, and the energy behind the communication. Respond with intention and empathy.

2. Quantum Pause in Conflict:

When faced with conflicts, embrace the Quantum Pause. Before reacting impulsively, take a conscious breath. Allow the pause to create space for reflection, fostering a more intentional and harmonious response.

3. Gratitude Expression:

Express gratitude within your relationships. Regularly acknowledge and appreciate the positive qualities in others. Whether through words, gestures, or small acts of kindness, cultivate an atmosphere of gratitude that resonates in your interactions.

4. Shared Mindful Practices:
Explore shared mindful practices with loved ones. Whether it's a joint meditation, a mindful walk, or even preparing and enjoying a meal together with heightened awareness, these shared rituals deepen connections on a quantum level.

Quantum Backwards Review: Reflecting on Your Evolution

The Quantum Backwards Review is a unique journaling exercise that invites you to reflect on your evolutionary journey through Quantum Mindfulness. Unlike traditional journaling, this exercise starts from the present and works backward, offering a fresh perspective on your growth and experiences. Follow these steps to engage in this insightful practice:

1. Present Reflection:

Begin your journal entry with a reflection on your current state of being. Consider your current mindset, emotions, and overall well-being. Observe the qualities and aspects of your consciousness that are most prominent in the present moment.

2. Recent Milestones:
Move backward in time and recall recent milestones or significant events in your life. Reflect on how these experiences have shaped your thoughts, emotions, and perspective. Consider both challenges and triumphs, acknowledging the lessons embedded in each.

3. Evolutionary Insights:
Continue moving backward through your memories, exploring the evolution of your consciousness. Identify key insights, realizations, or shifts in perception that have marked your journey. Observe the quantum leaps in your awareness and the transformative moments that have contributed to your growth.

4. Mindful Practices:
Examine the mindful practices and Quantum Mindfulness techniques you have incorporated into your routine. Reflect on how each practice has influenced your daily life, relationships, and overall well-being. Identify patterns of consistency and areas where you have experimented with new approaches.

5. Intentional Awareness:
Explore moments of intentional awareness that stand out in your backward journey. These could be instances where you consciously applied Quantum Mindfulness principles, experienced a heightened state of presence, or navigated challenges with a quantum-conscious perspective.

6. Gratitude for the Journey:
Express gratitude for the evolutionary journey you have undertaken. Acknowledge the effort and commitment you've invested in fostering

intentional awareness and self-discovery. Celebrate the progress you've made and the quantum-conscious shifts that have unfolded along the way.

7. Future Intentions:

Conclude your Quantum Backwards Review with future intentions. Consider the aspects of your consciousness you aim to further develop and the areas of life where you aspire to apply Quantum Mindfulness. This forward-looking reflection sets the stage for the next chapters of your quantum-conscious journey.

By engaging in the Quantum Backwards Review, you not only gain a retrospective view of your growth but also cultivate a deeper understanding of the interconnectedness of your experiences. This exercise serves as a unique tool for self-reflection, offering insights into the quantum dimensions of your consciousness and the continuous evolution that shapes your present and future self.

As you navigate the practical applications of Quantum Mindfulness, recognize that each exercise and real-life example is a stepping stone toward personal transformation. Through immediate implementation, you unlock the potential to elevate your consciousness, enhance well-being, and embark on a journey of self-mastery. May this chapter be a guide to seamlessly integrate Quantum Mindfulness into your life, creating a new way of mindful living that transcends the ordinary and propels you towards extraordinary possibilities.

Chapter 9: Quantum Mastery

Elevating Your Potential Beyond the Mind's Horizon:

In this concluding chapter, we unveil the pinnacle of Quantum Mindfulness: the journey toward Quantum Mastery. Beyond ordinary mindfulness, Quantum Mastery transcends the mind's conventional boundaries, inviting you to explore the limitless potential of your consciousness. Let's examine the transformative principles and practices that empower you to elevate your existence and embrace the quantum dimensions of your being.

Quantum Resilience: Navigating Life's Waves

Quantum Resilience is a state of being that transcends conventional notions of resilience. It is a dynamic and transformative approach to navigating the ebb and flow of life with grace and purpose. Imagine yourself as a surfer riding the waves of existence, not merely surviving but thriving in the face of challenges. Here, we give you practical tools and perspectives that foster Quantum Resilience, empowering you to embrace the ever-changing currents of life with a quantum-conscious resilience.

Tools for Quantum Resilience:

1. Mindful Adaptability:
Embrace the art of mindful adaptability, recognizing that change is a constant in life. Quantum Resilience involves staying present and adaptable in the face of unexpected shifts. Practice mindfulness to navigate transitions with greater ease and openness.

2. Quantum Perspective Shifts:
Cultivate the ability to shift perspectives, seeing challenges not as obstacles but as opportunities

for growth. Quantum Resilience involves viewing setbacks from a broader lens, extracting lessons, and transforming adversity into catalysts for personal and spiritual evolution.

3. Conscious Breathwork:
Utilize conscious breathwork as a tool to anchor yourself in the present moment. Quantum Resilience thrives on the ability to stay centered amidst life's storms. Incorporate breathing exercises to calm the mind, foster clarity, and build inner strength.

4. Quantum Affirmations:
Craft Quantum Affirmations that resonate with resilience and empowerment. These affirmations, rooted in the principles of quantum consciousness, reinforce your ability to navigate challenges with strength and purpose. Repeat them regularly to reinforce a resilient mindset.

Perspectives for Quantum Resilience:

1. Embracing Impermanence:
Acknowledge the impermanence of life. Quantum Resilience involves understanding that circumstances, both favorable and challenging, are temporary. Embracing impermanence allows you to flow with life's changes rather than resist them.

2. Trusting the Process:
Cultivate trust in the unfolding process of life. Quantum Resilience requires surrendering to the innate intelligence of the universe. Trust that challenges are part of a larger plan, guiding you toward greater self-discovery and evolution.

3. Integrating Quantum Mindfulness:
Infuse Quantum Mindfulness into your daily life. Quantum Resilience is nurtured by the intentional awareness that comes with mindfulness practices. Stay present, observe your thoughts, and respond to challenges with a clear and focused mind.

4. Quantum Self-Compassion:
Extend compassion to yourself as you navigate life's waves. Quantum Resilience involves acknowledging your vulnerabilities without judgment. Practice self-compassion, recognizing that setbacks are not reflections of inadequacy but opportunities for learning and growth.

Quantum Resilience is not about avoiding challenges but about navigating them with conscious intention and grace. By incorporating these tools and perspectives into your life, you can cultivate a resilience that goes beyond mere survival, allowing you to thrive in the ever-changing currents of existence.

Quantum Alignment: Harmonizing with Universal Flow

This section explores the art of aligning your consciousness with the cosmic rhythm, unlocking the doors to synchronicity, abundance, and a profound sense of purpose.

Quantum Alignment is not merely a mental exercise; it is a state of being that resonates with the very fabric of the universe.

Practices for Quantum Alignment:

1. Intentional Alignment Ritual:
Create a ritual for intentional alignment, a dedicated space and time where you consciously attune your energy to the universal flow. This may involve meditation, visualization, or simply grounding exercises that connect you with the present moment and the expansive energy of the cosmos.

2. Quantum Manifestation Techniques:
Explore Quantum Manifestation, a practice that goes beyond traditional manifestation. Quantum Alignment involves aligning your desires with the vibrational frequency of the universe. Engage in visualization, affirmations, and creative expression to manifest your intentions in harmony with the cosmic flow.

3. Mindful Energy Flow:
Practice mindful awareness of your energy flow throughout the day. Quantum Alignment thrives on conscious energy management. Regularly check in with yourself, ensuring that your thoughts, emotions, and actions are in alignment with your higher intentions and the universal energy that surrounds you.

4. Quantum Rituals of Connection:
Engage in rituals that deepen your connection with the universe. Whether it's stargazing, communing with nature, or engaging in sacred practices from various wisdom traditions, Quantum Alignment involves recognizing the interconnectedness of all things and aligning your consciousness with this cosmic tapestry.

Perspectives for Quantum Alignment:

1. Trusting the Cosmic Symphony:

Develop trust in the cosmic symphony, understanding that there is an inherent order and intelligence in the universe. Quantum Alignment requires surrendering to the rhythm of life, trusting that your journey is part of a grander orchestration.

2. Aligning with Universal Principles:
Explore universal principles such as love, compassion, and gratitude. Quantum Alignment involves aligning your consciousness with these timeless principles, recognizing them as the guiding forces that weave through the fabric of the cosmos.

3. Quantum Connection Meditation:
Practice the Quantum Connection Meditation, a contemplative journey where you envision yourself as an integral part of the vast cosmic web. Quantum Alignment involves feeling the interconnectedness with all of existence, fostering a sense of unity and purpose.

4. Quantum Flow States:
Embrace states of flow in your daily life.
Quantum Alignment is facilitated by moments
where you effortlessly engage in activities,
losing track of time and ego. Recognize these
flow states as indications that you are in
harmony with the universal flow.

Quantum Alignment is an ongoing exploration, a
conscious journey where you attune your being
to the universal frequency. By incorporating
these practices and perspectives into your life,
you embark on a transformative path of
harmonizing with the cosmic dance, unlocking
the boundless possibilities that arise when your
consciousness aligns with the universal flow.

Quantum Creativity: Tapping into the Source of
Innovation

Quantum Creativity emerges as a boundless
wellspring, inviting you to tap into the source of
innovation that transcends conventional

thinking. This section offers a comprehensive exploration of practices, perspectives, and principles that stimulate creative insights, encourage nonlinear thinking, and connect you to the wellspring of inspiration inherent in the quantum dimensions of your consciousness.

Practices for Quantum Creativity:

1. Quantum Imagination Activation:
Engage in Quantum Imagination Activation, a practice that transcends traditional boundaries of imagination. Allow your mind to wander freely, exploring areas beyond the known. This practice encourages the generation of novel ideas by tapping into the limitless possibilities existing in the quantum field of your consciousness.

2. Quantum Flow State Creation:
Cultivate Quantum Flow States as a foundation for creative expression. These states involve a seamless immersion in activities where time

seems to vanish, and inspiration effortlessly flows. Recognize and embrace these moments, as they signify a harmonious connection with the quantum dimensions of your creative potential.

3. Quantum Synergy Collaboration:
Explore Quantum Synergy Collaboration, a practice that involves connecting with others to co-create innovative solutions. Quantum Creativity thrives in collaborative spaces where diverse perspectives converge, creating a dynamic synergy that goes beyond individual contributions.

4. Quantum Playfulness Integration:
Integrate Quantum Playfulness into your creative process. Embrace a childlike curiosity and playfulness, allowing your mind to explore without the constraints of judgment or preconceived notions. Quantum Creativity flourishes in an atmosphere of lighthearted exploration.

Perspectives for Quantum Creativity:

1. Nonlinear Thinking Embrace:
Nonlinear Thinking becomes a cornerstone of Quantum Creativity. Release linear constraints and welcome the nonsequential, unconventional, and unpredictable nature of creative insights. Embrace the idea that breakthroughs often come from unexpected and nonlinear pathways.

2. Quantum Observation Mindset:
Adopt a Quantum Observation Mindset, where you become a keen observer of the world around you. Notice patterns, connections, and subtle nuances that may elude casual observation. Quantum Creativity involves drawing inspiration from the intricate tapestry of existence.

3. Quantum Resonance Connection:
Connect with Quantum Resonance, recognizing that creative ideas resonate with the vibrational

frequency of your consciousness. Trust your intuitive sense of what feels resonant and authentic. Quantum Creativity involves tuning into the frequencies that align with your unique creative expression.

4. Quantum Limitless Possibility Belief: Cultivate a belief in Quantum Limitless Possibility. Quantum Creativity requires letting go of limiting beliefs and embracing the idea that the quantum dimensions of your consciousness hold an infinite reservoir of creative potential. Trust that innovative solutions can emerge from the uncharted realms of your mind.

Quantum Creativity is not a linear process but a dynamic dance with the infinite possibilities residing within your consciousness. By integrating these practices and perspectives into your creative endeavors, you unlock the doors to a quantum field of inspiration, innovation, and boundless expression. Embrace the fluidity of

Quantum Creativity, allowing it to guide you into unexplored territories of imaginative brilliance.

Quantum Presence: Embracing the Eternal Now

This section explores the art of embracing the Eternal Now, where past and future dissolve, leaving only the profound immediacy of the present. Quantum Presence is not a fleeting state but a way of being that amplifies your awareness, allowing you to experience life with heightened clarity, appreciation, and a deep sense of interconnectedness.

Practices for Quantum Presence:

1. Mindful Breath Anchoring:
Anchor yourself in the present moment through Mindful Breath Anchoring. This practice involves focusing your attention on each breath, allowing it to draw you into the essence of the Eternal Now. Feel the sensations of each inhalation and

exhalation, fostering a profound sense of presence.

2. Quantum Sensory Awareness:
Engage in Quantum Sensory Awareness, where you consciously attune your senses to the immediate surroundings. Embrace the sights, sounds, smells, tastes, and tactile sensations that envelop you. Quantum Presence involves fully immersing yourself in the sensory tapestry of the present.

3. Quantum Non-Judgmental Observation:
Practice Quantum Non-Judgmental Observation, where you observe thoughts, emotions, and sensations without attaching judgment. Embracing the Eternal Now requires cultivating a non-reactive awareness, allowing each moment to unfold without the interference of preconceived notions.

4. Quantum Stillness Meditation:

Embark on Quantum Stillness Meditation, a practice that leads you into the depths of inner silence. Allow the stillness to envelop your consciousness, transcending the boundaries of time. Quantum Presence is cultivated in the silent expanses of the mind.

Perspectives for Quantum Presence:

1. Surrendering to the Present Flow:
Embrace the art of surrendering to the Present Flow. Quantum Presence involves letting go of the incessant need to control or anticipate the future. Surrendering to the flow allows you to ride the currents of the Eternal Now with grace and acceptance.

2. Quantum Gratitude Cultivation:
Cultivate Quantum Gratitude as a lens through which you perceive the present. Express gratitude for the simple and profound aspects of the current moment. Quantum Presence is

heightened when gratitude becomes a natural response to the richness of your experience.

3. Quantum Timelessness Recognition: Recognize the timelessness inherent in every moment. Quantum Presence involves understanding that time is a construct of the mind, and true presence resides in the eternal nature of now. Release the illusion of past and future, anchoring yourself in the perpetual now.

4. Quantum Interconnected Awareness: Develop an awareness of Quantum Interconnectedness. Realize that your presence is not isolated but woven into the fabric of all existence. Quantum Presence involves recognizing the interconnected dance of life and embracing your role in this cosmic symphony.

Quantum Presence is a state of being that transcends the limitations of time, offering you a profound connection to the boundless dimensions of your consciousness. By incorporating these practices and perspectives into your daily life, you open the door to a deeper, more meaningful experience of the Eternal Now, allowing each moment to unfold with clarity, purpose, and a sense of awe-inspiring interconnectedness.

Quantum Integration: Weaving Mindfulness into Being

As we continue with Quantum Mastery, Quantum Integration emerges as the art of seamlessly weaving mindfulness into the very fabric of your being. This section explores practices, perspectives, and principles that guide you in incorporating Quantum Mindfulness into every aspect of your existence. Quantum Integration is not a compartmentalized endeavor but a holistic approach, inviting you to live in a state of intentional awareness that permeates your thoughts, emotions, actions, and relationships.

Practices for Quantum Integration:

1. Quantum Daily Mindfulness Ritual:
Establish a Quantum Daily Mindfulness Ritual, a dedicated time each day where you consciously engage in mindfulness practices. This may include meditation, breathwork, or reflective exercises that set the tone for intentional awareness throughout the day.

2. Quantum Mindful Reflections:
Incorporate Quantum Mindful Reflections into your daily routine. Take moments to pause, reflect, and bring mindful awareness to your thoughts and emotions. Quantum Integration involves consistently checking in with yourself, fostering a continuous state of self-awareness.

3. Quantum Intention Setting:
Practice Quantum Intention Setting, where you align your daily intentions with the principles of mindfulness. Begin each day with a clear intention to live with awareness, compassion,

and presence. Quantum Integration is nurtured when your intentions become the guiding compass for your actions.

4. Quantum Mindful Movement:
Engage in Quantum Mindful Movement, incorporating awareness into your physical activities. Whether it's walking, yoga, or other forms of exercise, allow each movement to be imbued with mindfulness. Quantum Integration involves synchronizing your physical and mental states in a harmonious dance.

Perspectives for Quantum Integration:

1. Holistic Awareness Embrace:
Embrace a Holistic Awareness perspective. Quantum Integration involves seeing mindfulness as a holistic approach that permeates all aspects of your life. Recognize that every thought, emotion, and action can be infused with intentional awareness.

2. Quantum Presence in Relationships:
Extend Quantum Presence into your relationships. Quantum Integration involves bringing mindfulness into your interactions, listening with full attention, and responding with conscious intention. Cultivate an environment of shared awareness and connection.

3. Quantum Emotional Regulation:
Explore Quantum Emotional Regulation, where mindfulness becomes a tool for navigating emotions. Rather than reacting impulsively, Quantum Integration involves responding to emotions with a conscious and measured awareness, fostering emotional resilience.

4. Quantum Gratitude as a State of Being:
Cultivate Quantum Gratitude as a state of being. Quantum Integration involves living with a continual sense of gratitude for the present moment and the interconnected web of existence. Allow gratitude to become an integral part of your consciousness.

Quantum Integration is an ongoing process, a journey where mindfulness becomes more than a practice—it becomes a way of being. By incorporating these practices and perspectives into your life, you embark on a transformative path of intentional awareness that seamlessly integrates with your thoughts, emotions, and actions. Quantum Integration is an invitation to live in a perpetual state of mindfulness, unlocking the full spectrum of your existence with clarity, compassion, and purpose.

Quantum Legacy: Shaping a Conscious Future

Quantum Legacy is the conscious shaping of a future deeply rooted in mindfulness, compassion, and intentional awareness. This section invites you to consider the imprint you leave on the world, the relationships you cultivate, and the impact of your consciousness on the unfolding story of humanity.

Practices for Shaping a Quantum Legacy:

1. Quantum Conscious Decision-Making:
Engage in Quantum Conscious Decision-Making, where each choice is guided by mindfulness and consideration of its broader implications. Quantum Legacy involves recognizing the interconnectedness of decisions and their potential to shape the course of your life and beyond.

2. Quantum Compassionate Leadership:
Embrace Quantum Compassionate Leadership in your personal and professional spheres. Quantum Legacy is nurtured when your actions are driven by empathy, understanding, and a commitment to collective well-being. Lead with a mindful awareness of the impact you have on others.

3. Quantum Eco-Mindfulness:
Practice Quantum Eco-Mindfulness, fostering an awareness of your ecological footprint. Quantum

Legacy involves considering the environmental impact of your choices and actively contributing to a sustainable and mindful relationship with the planet.

4. Quantum Generosity and Service:
Incorporate Quantum Generosity and Service into your life. Quantum Legacy is shaped by the positive contributions you make to the lives of others. Actively seek opportunities to serve, share, and uplift those around you with a mindful and compassionate heart.

Perspectives for Shaping a Quantum Legacy:

1. Quantum Ripple Effect Recognition:
Acknowledge the Quantum Ripple Effect of your actions. Quantum Legacy involves understanding that every thought, word, and deed sends ripples into the fabric of existence, influencing not only your immediate surroundings but the broader collective consciousness.

2. Quantum Interconnected Co-Creation:
Embrace the concept of Quantum Interconnected Co-Creation. Quantum Legacy is forged in collaboration with the world around you. Recognize that your contributions are part of a collective dance, co-creating the future with a shared vision of mindfulness, harmony, and understanding.

3. Quantum Educational Mindfulness:
Infuse Quantum Educational Mindfulness into your interactions with knowledge and learning. Quantum Legacy involves recognizing the transformative power of education. Cultivate a mindful approach to acquiring and sharing knowledge, fostering a conscious evolution of understanding.

4. Quantum Intergenerational Connection:
Nurture Quantum Intergenerational Connection. Quantum Legacy extends beyond individual lifetimes. Cultivate relationships and share

wisdom across generations, creating a legacy that transcends time and contributes to the ongoing evolution of consciousness.

Shaping a Quantum Legacy is a profound responsibility and a testament to the conscious evolution of humanity. By incorporating these practices and perspectives into your life, you become an active participant in the co-creation of a future that reverberates with mindfulness, compassion, and the enduring impact of intentional awareness. Your Quantum Legacy is not just a personal narrative; it is an integral thread in the vast tapestry of human consciousness, weaving a story of interconnected evolution and mindful living.

Quantum Evolvement: A Continuous Journey

Quantum Evolvement is a perpetual journey of self-discovery, mindfulness, and the continuous expansion of consciousness. In this section you will embrace the idea that growth is a lifelong process, and each moment offers an opportunity for evolution, transformation, and the deepening of your connection to the quantum dimensions of your being.

Practices for Continuous Quantum Evolvement:

1. Quantum Reflective Awareness:
Cultivate Quantum Reflective Awareness, where you regularly pause to reflect on your experiences, choices, and the evolving landscape of your consciousness. Quantum Evolvement involves an ongoing dialogue with your inner self, fostering self-awareness and personal growth.

2. Quantum Learning and Adaptability:
Embrace Quantum Learning and Adaptability as foundational principles. Quantum Evolvement thrives on a willingness to explore new ideas, adapt to change, and approach life with a curious and open mind. Every experience becomes a stepping stone for further growth.

3. Quantum Resilience Building:
Develop Quantum Resilience as a core attribute. Quantum Evolvement involves navigating the inevitable challenges of life with a resilient spirit, viewing setbacks as opportunities for learning and transformation. Strengthen your inner resilience through mindful responses to adversity.

4. Quantum Compassionate Self-Exploration:
Engage in Quantum Compassionate Self-Exploration, where you approach your own inner landscape with kindness and understanding. Quantum Evolvement requires a deep connection with your authentic self, embracing both strengths and vulnerabilities with compassion.

Perspectives for Continuous Quantum Evolvement:

1. Embracing the Uncharted:

Embrace the Uncharted territories of your consciousness. Quantum Evolvement involves stepping into the unknown with courage and curiosity, recognizing that true growth often occurs beyond the familiar boundaries of comfort.

2. Quantum Flow with Life's Rhythms:
Flow with Life's Rhythms in a state of Quantum Acceptance. Embrace the natural cycles of growth, transformation, and renewal. Quantum Evolvement involves aligning with the ebb and flow of life, acknowledging that every phase holds valuable lessons.

3. Quantum Connectivity to Universal Consciousness:
Deepen your Quantum Connectivity to Universal Consciousness. Recognize that the evolution of your consciousness is intricately woven into the fabric of the cosmos. Quantum Evolvement involves realizing your interconnectedness with

all of existence, fostering a sense of unity and purpose.

4. Quantum Contribution Mindset:
Adopt a Quantum Contribution Mindset. Quantum Evolvement is not only about personal growth but also about contributing to the well-being of the collective. Consider how your evolving consciousness can positively influence the world around you.

Quantum Evolvement is an inclusive and expansive journey that transcends individual boundaries. By incorporating these practices and perspectives into your life, you not only deepen your personal evolution but also contribute to the broader tapestry of human consciousness. May the continuous exploration of the quantum dimensions within lead you to a life rich in mindfulness, purpose, and an ever-unfolding path of self-discovery. May this journey of Quantum Evolvement inspire and help as many people as possible on their unique paths of growth and transformation.

In the pursuit of Quantum Mastery, you elevate your potential beyond the mind's horizon, tapping into the infinite possibilities that reside within your consciousness. This chapter serves as a guide, offering insights, practices, and principles that empower you to embrace the full spectrum of your existence and unlock the quantum dimensions of your mastery. May your journey toward Quantum Mastery be a transformative odyssey, illuminating the path to self-realization and boundless potential.

Conclusion: Nurturing Quantum Mindfulness

As we conclude this exploration of Quantum Mindfulness, let's reflect on the transformative journey we've undertaken. We've explored the interconnected dance of quantum principles and mindfulness, unlocking the potential for profound self-discovery, intentional awareness, and continuous evolution. The key takeaways from this exploration serve as guiding lights, inviting you to infuse mindfulness into every facet of your life and contribute to the collective tapestry of human consciousness.

Key Takeaways:

1. Embracing Quantum Awareness:
Cultivate an awareness that transcends the surface of everyday experiences. By acknowledging the interconnectedness of

quantum principles and mindfulness, you unlock the door to a deeper understanding of yourself and your place in the grand symphony of existence.

2. Living in the Eternal Now:

The Eternal Now is not a fleeting concept but a profound way of being. Embrace the present moment with a mindful heart, recognizing that the past and future are threads woven into the timeless tapestry of now. In this state, you discover the richness of life, unburdened by the weight of yesterday or the uncertainties of tomorrow.

3. Shaping a Quantum Legacy:

Your journey is not just an individual narrative but a contribution to the collective consciousness. With every intentional choice, compassionate interaction, and mindful decision, you shape a Quantum Legacy that transcends your lifetime, leaving an imprint of awareness, compassion, and conscious living for generations to come.

4. Quantum Resilience and Adaptability:
Embrace challenges with a Quantum Resilience mindset, recognizing that adversity is an inherent part of life's journey.
Cultivate Quantum Adaptability, allowing your mindset to evolve and adjust in response to changing circumstances.

5. Quantum Gratitude and Appreciation:
Foster a mindset of Quantum Gratitude, appreciating the abundance present in your life and the interconnectedness with the larger web of existence.
Practice Quantum Appreciation for the simple joys and experiences, recognizing the beauty in the ordinary moments of life.

6. Quantum Compassion and Connection:
Infuse Quantum Compassion into your interactions, understanding the shared human

experience and offering empathy to those around you.

Cultivate Quantum Connection by recognizing the threads of interdependence that weave through your relationships, fostering a sense of unity.

7. Quantum Playfulness and Creativity:

Approach life with a sense of Quantum Playfulness, allowing your inner child to explore, create, and find joy in the spontaneity of the present moment.

Channel Quantum Creativity by tapping into your imaginative potential, exploring novel ideas, and embracing innovative thinking.

8. Quantum Stillness and Reflection:

Regularly engage in Quantum Stillness practices, allowing moments of inner silence to deepen your connection to the quantum dimensions within.

Embrace Quantum Reflective Awareness as a tool for continuous self-discovery, using reflection as a means to understand and learn from your experiences.

9. Quantum Harmony and Balance:
Seek Quantum Harmony by finding balance in various aspects of your life, acknowledging the interconnected nature of mind, body, and spirit. Explore Quantum Equanimity, cultivating a calm and balanced mind that remains centered amidst life's fluctuations.

10. Quantum Intention and Mindful Action:
Set Quantum Intentions that align with your values, guiding your actions with conscious purpose and direction.
Practice Quantum Mindful Action by approaching tasks with full attention, bringing intentionality to even the simplest of activities.

11. Quantum Silence and Communication:
Value the power of Quantum Silence, recognizing that sometimes the most profound insights arise in moments of stillness.

Enhance Quantum Communication by fostering mindful and intentional dialogue, ensuring your words align with the principles of compassion and understanding.

12. Quantum Joy and Celebration:

Invite Quantum Joy into your life, finding delight in both the extraordinary and the ordinary moments. Practice Quantum Celebration, acknowledging milestones, achievements, and the beauty of the journey with a heart full of gratitude.

Continued Exploration: Daily Examples for Maximum Benefit:

Morning Quantum Mindfulness Ritual:

Begin your day with a Quantum Daily Mindfulness Ritual. Engage in mindful breathing, setting clear intentions for the day ahead.

Practice Quantum Imagination Activation during your morning routine, allowing your mind to explore creative possibilities for the day.

Throughout the day, practice Quantum Presence by anchoring yourself in the present moment during routine activities, such as sipping your morning coffee or taking a shower.

Afternoon Quantum Integration:

During your workday, practice Quantum Mindful Reflections by taking short breaks to reflect on your thoughts and emotions.

Infuse Quantum Mindful Movement into your lunch break, bringing mindful awareness to each bite during a mindful eating exercise.

Engage in Quantum Generosity and Service by offering a helping hand to a colleague or expressing gratitude for the shared moments.

Evening Quantum Legacy Building:

Practice Quantum Reflective Awareness in the evening, reflecting on your experiences and choices throughout the day.

Embrace Quantum Learning and Adaptability by engaging in an evening activity that introduces a new skill or perspective.

Before bedtime, engage in a Quantum Stillness Meditation, allowing your mind to settle into a state of tranquility, preparing you for a restful sleep.

As we conclude this exploration, let these daily examples be the catalysts for a continuous journey of Quantum Mindfulness. May each intentional breath, compassionate interaction, and mindful choice contribute to the ongoing evolution of your consciousness and the collective consciousness of humanity. May the tapestry of your life be woven with threads of awareness, purpose, and the boundless possibilities found within the quantum dimensions of your being. Continue this journey with an open heart, a curious mind, and the unwavering belief that the exploration of Quantum Mindfulness is a lifelong adventure of self-discovery and transformative growth.

With infinite love,

Harmony Weaver

www.ingramcontent.com/pod-product-compliance
Lightning Source LLC
Chambersburg PA
CBHW070859290526
45795CB00001B/177